W9-BBA-477

A Cash-Free Society

A Cash-Free Society

Whether We Like It or Not

Kai A. Olsen

ROWMAN & LITTLEFIELD
Lanham • Boulder • New York • London

Published by Rowman & Littlefield
An imprint of The Rowman & Littlefield Publishing Group, Inc.
4501 Forbes Boulevard, Suite 200, Lanham, Maryland 20706
www.rowman.com

Unit A, Whitacre Mews, 26-34 Stannary Street, London SE11 4AB

The author has received support from Norwegian Nonfiction Writers and Translators
Association (NFF)

British Library Cataloguing in Publication Information Available

Library of Congress Cataloging-in-Publication Data
Names: Olsen, Kai A., author.
Title: A cash-free society : whether we like it or not / Kai A. Olsen.
Description: Lanham : Rowman & Littlefield, [2018] | Includes bibliographical
   references and index.
Identifiers: LCCN 2018012681 (print) | LCCN 2018013507 (ebook) | ISBN
   9781442227439 (electronic) | ISBN 9781442227422 (cloth : alk. paper)
Subjects: LCSH: Money. | Cash transactions. | Electronic funds transfers. |
   Technological innovations—Social aspects. | Technological
   innovations—Economic aspects.
Classification: LCC HG221 (ebook) | LCC HG221 .O47 2018 (print) | DDC
   332.4—dc23
LC record available at https://lccn.loc.gov/2018012681

♾ The paper used in this publication meets the minimum requirements of American
National Standard for Information Sciences—Permanence of Paper for Printed Library
Materials, ANSI/NISO Z39.48-1992.

Printed in the United States of America

Contents

Preface   vii

1   Moving to a Digital World   1

Part One   Computer Applications   7

2   Computer Applications   9

3   Complex Computer Applications   25

Part Two   The Cash-Free Society   41

4   Money   43

5   Uncle Joe's Island   53

6   From Analog to Digital   57

7   Fundamentals for a Digital Economy   65

8   Infrastructure for Digital Payments   71

9   Digital Payments   81

10   Internet Banks   87

11   Virtual Currencies   93

12   Advantages of a Digital Payment System   101

13   Disadvantages of a Digital Payment System   111

14   Case: Norway   123

15   New Systems   141

16   The Cash-Free society   147

Index   153

About the Author   157

Preface

We are moving into a digital world. Many of the operations that were previously performed manually on paper are now being handled by a computer. Now, as we look into the future, we can talk about smart algorithms, artificial intelligence, big data, and machine learning and discuss the possibility of applications such as autonomous cars, decision making machines, and smart robots. While the prospects are fascinating and perhaps also a bit scary, this book focuses on the impact of a more prosaic technology—digital payments.

Digital payments do not require any breakthroughs in technology; the technology is already here. The terminals, the Internet banking, the regulations, and all the software are in place. In some countries nearly all payments are performed digitally. Cash is being marginalized. With new and even simpler payment systems, such as smartphones and tap-to-pay credit cards, it is realistic to expect that some countries will be practically cash-free in a few years. This has implications for consumers, merchants, banks, and for society as a whole.

A central part of this book is the discussion of these implications. They will require a small change for some, a large change for others. Some advantages of cash will be lost, but, as we shall see, the digital solutions can offer great advantages for all, with the exception of those who participate in the black economy. This is because digital transactions are normally traceable, which implies that they can be checked to see if they are "white," or legal. Because cash is "anonymous," it can be used for transactions of all colors: white, gray or black. Criminals—such as thieves, dealers in narcotics, tax evaders, employers that do not follow regulations—will all have problems when they cannot participate in the normal economy.

This book is in two parts. The first presents the fundamentals of new computer applications, defining the background for studying digital payment systems that are introduced in part two. With the hype that we see today around new applications for computers, a good understanding of the possibilities and limitations of information technology is paramount. Part one consists of chapters 2 and 3. Part two starts with Chapter 4.

After the introduction (Chapter 1), Chapter 2 provides a detailed discussion of the tasks that invite a computer takeover, and those that are more difficult. Chapter 3 sets a frame of reference by discussing complex computer applications. These applications will be very difficult to turn into working systems, but are also very different from what we need in order to implement the cash-free society.

In Chapter 4 I discuss the various aspects of money. Whether represented as cash or as bits in a computer, money is an important ingredient for a working economy, for storing wealth, as a means of valuation, and for performing payments. In Chapter 5 I discuss the transition from analog to digital, from checks and credit cards based on paper to the digital versions. The fundamentals of a digital economy are presented in Chapter 6.

A digital society needs an infrastructure to work. This consists of point-of-sale terminals, smartphones, computer networks, clearing systems for transactions, and various digital payment systems. A detailed overview of this infrastructure is provided in Chapter 7, before Chapter 8 concentrates on the actual digital payments.

Internet banks are important ingredients in a digital economy; these are discussed in Chapter 9. True digital currencies such as bitcoin, which only exist in a digital form, are presented in Chapter 10. In Chapter 11 I give an introduction to seigniorage—that is, the income central banks acquire by printing money, along with some other important concepts.

Advantages of a digital payment system will be discussed in Chapter 12, the disadvantages in Chapter 13. In Chapter 14 I use Norway as a case. Here cash only accounts for 3 percent of all transactions today—a percentage that is reduced every year. Norway, along with Iceland, Denmark and Sweden will, in practice, be cash-free in a few years. Interestingly, this is a consumer-led evolution. Consumers want simple and secure systems for payments, and banks and other financial companies are providing the infrastructure. The central banks and the politicians are not engaged. As we shall see, this may be a problem, but societies will be cash-free whether we like it or not.

All countries are moving in this direction, albeit at a different rate. Smaller, homogeneous countries have an advantage over larger countries in this regard, as experiments can be carried out more easily, and the cost of implementing new systems is much lower. Unsurprisingly, Scandinavia leads the way in terms of digitalization of payments. A criminal trying to rob a Norwegian bank would be disappointed as there would be no cash! Nearly all payments are digital and fewer and fewer people carry cash in their wallet. In this respect, these countries offer a peek into the future. What is happening there today will be the norm in other countries within only a few years.

Chapter 15 explores how new systems can make digital payments even simpler than they are today. I also discuss how the data generated from the digital economy can be useful for the authorities as well as private companies and ordinary citizens. Chapter 16 sums everything up.

Kai A. Olsen, kai.olsen@himolde.no
Professor in informatics, Molde University College, Molde, Norway
Professor II, Department of Informatics, University of Bergen, Norway
Professor II, Oslo Metropolitan University, Norway
Adjunct professor, School of Computing and Information, University of Pittsburgh

Chapter 1
Moving to a Digital World

In 1915, the chancellors of a town in Norway passed a regulation stating that trucks should not go faster than 15 km per hour. This was not a safety measure; instead, the chancellors wanted to protect the market for horse-drawn wagons, which had this as their maximum speed. Of course, the regulation did not work. The trucks went faster and would soon dominate freight transport in the city.

The traditional model for the sale of music was albums: a collection of several tracks on the same vinyl record. Later, cassette tapes were also used. Then the record was offered as a compact disk. These changes of technology were sustainable for the industry. If the record shop sold a vinyl record, a cassette tape, or a CD, it was the same business model. The disruption came when "pirates" offered music for downloading or streaming online. Clearly, this was a much better model for most music lovers than retaining the physical media. On the Internet, one could have large archives of music, good search and recommendation systems, and stream to many different types of devices. However, the music industry wanted to retain the model that gave them a secure income; they fought back, but couldn't win. The main thing was not that the pirates offered free music, but that it was much more convenient for users to get the music directly online. We see this today. When the industry is at last giving up on the physical media, a large number of customers are willing to pay for various added services on the new legal sites. For the most part the pirates have gone away— their job has been done.

These two very different examples show us that it is difficult to stop a new technology that has clear advantages. One may delay the introduction, but if the advantages are great enough, a momentum will build up that cannot be stopped. For music, this momentum was built when more and more people had access to the Internet, with increased bandwidth and inexpensive, portable digital music players. The nail in the coffin was the smartphone. When this became the unit of choice for playing music, there was no way back. In retrospect, we see that music on any type of physical media would have to be abandoned as the smartphone became the music player of choice. That is, in the first phase the alternative technology often comes as a replacement; in our example replacing physical media by streaming. In the next phase, a consumer explores the freedom that he or she gets when leaving the constraints of physical media. With every new breakthrough technology, some users retain the old. For music we see a thriving, albeit small, market for vinyl records. Some photographers, both amateurs and professionals, still use traditional film cameras. However, in this book we look at major trends, accepting that there always will be smaller niches that continue using a traditional technology.

While the record industry managed to maintain its business model for a few years, it may be argued that they would have been better off if they had taken control over the new technology instead of trying to stop it. At least they would then have avoided coping with the free services that the pirates provided. To make an impact after the pirates, this implied that they had to offer a free option when introducing their new services.

We often see that newcomers in an area have a better understanding of the potential of a new technology than the incumbents. This is probably because the incumbents, such as the record industry, are comfortable in a model that has served them for many years. It is difficult, and sometimes impossible, to break out. A good example is Kodak, which invented the digital camera but was too entrenched in its very successful business model of selling photographic film and analog cameras to exploit any other alternatives.[1] The companies that sold expensive mainframe computers had a business model that could not incorporate new mini-computers.[2] Later on, the same thing happened with the successful mini-computer manufacturers. Their organization was not able to design, manufacture, and sell the new PCs. For example, their salespeople were paid in the form of a percentage of the contract. This would give them a good bonus when the contracts were in the millions of dollars, but not when a customer bought a PC for less than a thousand dollars. Furthermore, while the mainframe and mini-computer manufacturers made the complete product, PC components and software are manufactured by many different companies. Therefore, a PC manufacturer requires high volumes to make money, which is quite different from the earlier technology.

These technology shifts have been so disruptive that one could argue that the incumbents would not have had a chance, even if they had decided to go for the new technology. It may be that Kodak's only choice was to continue making photographic film and analog cameras as long as it was profitable, and then close down its entire operation. The new digital age required different technologies, different competence, a different way of making and selling cameras, and different media for storing the images, and maybe Kodak would not have had a chance in this business? When smartphones became the standard ways of taking photographs, the consequence is straightforward—Kodak would be lost![3] Of course, shareholders in Kodak would have had the option to sell the shares and invest in companies that utilized the digital technology. But moving an investment is simpler than reorganizing a company from one technology platform to another disruptive platform.

Before the Internet, the major airlines had built a system where tickets were sold through travel agencies or airline offices. A customer would call in, describe his or her needs, and an operator would enter the booking in the system, then send the tickets by mail. With skilled operators, one could have a complex discount system, such as a discount on round trips or a discount for couples whereby if one paid the full price the other would pay half price.

With the advent of the Internet, all airlines, both the incumbents and the new low-cost airlines, established online booking systems, thereby allowing customers direct access to their systems. In this respect the technology was not disruptive; it was sustainable, for both the customer and the airline, but perhaps not for the intermediaries—the travel agencies. However, while many of the incumbents retained their complex discount structure, budget airlines such as Ryanair and Norwegian developed a much simpler discount system. The main idea was to separate tickets. Round-trip discounts and discounts for couples required two tickets to be connected. This could cause problems. For example, what would happen if only the full-price ticket for a couple was cancelled? Would this be possible or would the discount system demand that both are cancelled in this situation? As we can see, the idea of connecting two tickets increases the complexity, possibly beyond what normal users can handle.

The new airlines came in with many other ideas—low fares being the most important—but a major part of their success in establishing their new business was by simplifying their price structures. The enabler for these business models was the Internet and the simpler price structures made it possible for the customers to book tickets on their own. Still, it is not a disruptive technology for the airlines, as seen by the fact that both the traditional airlines (most of them anyway) and the low-cost airlines coexist. However, the Internet offered an opening for the new airlines. Another example of a sustainable technology in the airline business is the move from propeller airplanes to jets. Again, in this case, the technology shift does not disrupt the business model.

In this book I shall discuss digital currency and digital payments. The technological basis is inexpensive point-of-sale terminals, mobile technology, and encompassing computer networks. Interestingly, here we also have "pirates," manifested as a set of new cryptocurrencies that provide an alternative to the traditional currencies. These are maintained by smart algorithms instead of a central bank. They are still just in an experimental phase, perhaps also with an uncertain future. I shall return to cryptocurrencies later, but shall first concentrate on technology that enables us to make digital payments within the traditional currencies.

An important advantage of digital payments is that they can be embedded naturally in the buying process. Customers are there to purchase something and paying is a required part of most transactions. By using cash, the actual payment is a separate part of the transaction, most often performed manually. In a store, the cashier must tell the customer what to pay, take the cash, count it, and offer change in return. When paying with a card, the amount can be captured electronically. The advantage of a physical store may be that the payment goes faster when it is digital, especially with tap-to-pay solutions where the card or the smartphone only has to touch the terminal. On the Internet there will be no good alternative to paying digitally. The integration of payments in the whole process is also very much apparent when buying tickets, for theaters or concerts. The process may involve selecting shows, dates, and seating. Similarly, we use apps to buy tickets for buses and trains. These apps will help us find routes, schedules, and may even tell us when the bus will actually arrive. We see that payments are just a part of these processes, a part that cannot be separated out as a cash transaction.

Digital technology is disruptive for traditional payments such as cash or checks, but here the fight is not between companies. In practice, payment systems have to be general if they are to work. The customer clearly does not want one system for each store; neither does the merchant want one system for each type of credit card. That is, if many customers or companies accept one form of payment, soon all the others will have to follow. There will be no competitive advantage if any form of digital payment can be used in any store, but digital payment systems allow for lower cost and better service. In many ways this is like electricity; it is clearly an advantageous technology with many benefits, but since it is available to all stakeholders there is no competitive advantage between businesses within a country.

Network effects are important here. The advantage of a new technology increases with the number of users. The owner of the first telephone had no one to call; today we can call people all over the world. As more and more customers use digital payments as a default, merchants will have to provide the necessary equipment. Many stores used to have a sign on the door saying, "We accept credit cards"; today we may see warnings such as "Cash only—we do not take cards." In many countries such a sign will turn away most customers.

These network effects are very apparent with mobile pay systems that can handle person-to-person transactions, and in this respect can take over from cash. To aid the introduction of the new technology, and to get the network effects, the providers let you also send money to people that are not connected to the system. The receiver will then have a few days to register in the system; if it does not, the money is returned to the sender. For most customers this is an opportunity to install the new systems.

Between countries we see another picture. Some, such as the Nordic countries, have embraced digital payments. Checks are no longer used, cash transactions have been reduced, credit card transactions are digital, Internet banking is the norm, and invoices are sent electronically. New technology, not least smartphones, is handling more and more transactions. The advantage is an effective economy. Fewer and fewer resources are used for performing payments; at the same time, the customer, the business, and the authorities will get improved service and better overview. With digital payments, all operations can be connected to a given customer, which means a company will get valuable data about their customers and what they buy.

Other countries, such as the United States, are maintaining inefficient payment systems where paper is still the important ingredient—in the form of paper checks and paper money. Thus, the actual payment will in itself be more expensive. Another downside is less overview, for customers, businesses, and the authorities.

One could argue that cash has functioned very well for hundreds of years, so why change to another system? I shall return to this discussion later on, but going digital has many advantages, for customers, merchants, and for societies. Therefore, all countries will see an increase in digital payments. Furthermore, cash is becoming increasingly obsolete in a world where everything from invoices to salaries is paid digitally. Cash is losing momentum; it cannot be used for online shopping, and even if used in a physical store it creates a problem for the merchant, who must transfer the cash to the bank. The merchant also faces the problem that there may be few banks that handle cash. This is already the situation in several countries today.

Cheap, fast, and reliable computers are taking over many jobs and many tasks. Some experts estimate that, within a few years, 90 percent of today's jobs will be performed by a machine, a robot, or a computer algorithm. This shows a lack of understanding of computer technology. Independent of application, the computer needs a *formalization* of the task at hand. As the end result, data and the operations will be broken into zeros and ones inside the computer—that is, as a predetermined formula. This program or "formula" will then be executed as a deterministic operation. While smart software can automate some jobs, it cannot handle all.

Some tasks have already been formalized. This is especially the case for operations regarding money. From the very first banks, these processes—establishing an account, inserting money into the account, withdrawal, interest calculations, and so on—have followed strict procedures. Nobody would want a sloppy bank to take care of their money. In the 1950s and 1960s, when the first computers emerged, the job of the programmers was to convert these procedures from natural language into a computer language. In principle, this was an easy task as both descriptions were formalized, and many of the operations regarding accounts and payments have the advantage of being simple.

Other tasks are more difficult to formalize. Today, leading technological companies are working on autonomous vehicles—cars that can drive themselves on ordinary roads. If this is to be achievable, they will need to formalize the act of driving, the traffic situation and the roads. In practice, they need to devise systems that can allow the car to follow the road, stop at a red light, turn to the left or right, avoid pedestrians and other cars, and more. As human drivers, we often have to use our understanding of a traffic situation in order to do the right thing. It is very difficult to program this understanding into a computer—that is, to formalize the task. One may be able to do this for some types of situations, but in practice there will be an unlimited number of exceptions to handle.

A lot of hype is concentrated on many of these unformalized and difficult tasks, from autonomous cars and decision-making robots to natural language speech translation. It seems that the media discussion has largely overlooked the digitalization of payments, even if this is so much easier to achieve. For payments, there is no need for the new technology to "understand" the customers' intentions. Also, the technology that is needed to perform all types of payments—that is, to implement the cash-free society—is already here.

One reason for this paradox may be that there are some very dedicated supporters of cash. There are arguments that the poor, very young, or elderly will fall outside a digital system. While we may expect that privacy issues that emerge when we replace anonymous cash with traceable digital payments are up for discussion, we must also recognize that many stakeholders feel that their "business model" is threatened in a cash-free society. Anonymity is the all-important feature for criminals, tax evaders, and terrorists. From a technical perspective, however, or from a cost-benefit viewpoint, digitalizing payments are a clear win, but the aggressiveness in the discussion may have scared politicians, financial institutions, and many others from being involved.

There are several advantages to digital payments. They can be performed quickly and efficiently, there is no need to be physically present to perform the payment, and they can be performed everywhere using the customer's own computer or smartphone. While there are clear advantages for using digital payments for customers, merchants, and banks, the greatest benefit may come to the society as a whole. Cash handling is expensive, and it is not very environmentally friendly, as large amounts of cash have to be transported in armored trucks. As we have seen, it also supports black market transactions. I shall explore these issues and also study how citizens in many countries have made the transition from using cash to become real digital citizens.

While there may be law-abiding ordinary citizens who want to operate anonymously, doing so is not very practical in a modern society. One would not have the opportunity to perform online shopping or to get the discounts that are offered to app users, for example regarding public transportation, concert tickets, and grocery stores. In fact, a person wishing to retain full anonymity cannot use social networks or have a mobile phone and must be very careful using any service on the Internet. Still, many activities, such as crossing a border, renting a car, or checking in at a hotel, cannot be performed anonymously. In some countries, toll-booths are digital, meaning that anonymity is also lost. Data on these activities will be registered in computer systems. One can question whether it really is possible to operate anonymously in a modern society. So, instead of trying to avoid being traced, a better option may be to support laws and regulations that stop the misuse of the data that are collected.

The world will become cash-free, whether we like it or not. Some countries are nearly there, and others will follow.

Notes

[1] John J. Larish (2012) *Out of Focus: The Story of How Kodak Lost Its Direction*, Createspace Independent Publishing Platform.

[2] See Clayton Christensen (1997) *The Innovator's Dilemma*, Harvard Business Review Press.

[3] Scott Anthony (2016) "Kodak's Downfall Wasn't About Technology," *Harvard Business Review*, July 15. https://hbr.org/2016/07/kodaks-downfall-wasnt-about-technology

Part One
Computer Applications

This part offers a background to understand computer applications, allowing one to see beyond the hype.

Chapter 2
Computer Applications

In order to allow a computer to perform a task, the task needs to be formalized: that is, expressed in unambiguous terms as a program. The program will be represented as binary digits within the computer, as a sequence of zeros and ones, offering an unambiguous and exact description of what to do.

Many important tasks were formalized long before the advent of computers. Take banking as an example. When a customer inserted an amount into a bank account in the seventeenth century, the bank followed a strict procedure. The amount was received by the teller, counted, and registered to the correct account. The customer received a receipt as a confirmation that the transaction had been performed correctly. At the end of the day, the bank teller would count the money in his cash register and compare that amount with the insertions and withdrawals during the day.

Banks used some of the very first computers. Their applications were ideal for the new technology. Most routines in banks were formalized, which means they followed strict routines. The job of a programmer was to rewrite these rules from natural language, such as English, into COBOL, FORTRAN, or any other programming language. It was also important that bank transactions did not need large amounts of data. Even with the very limited first computers, one could perform useful operations, such as balancing accounts or computing interest.

The requirement for formalization is necessary for any kind of automation. If we go back to the industrial revolution, we see that early machines embodied the formalization of a task, whether it was pumping, knitting, spinning, weaving, or picking cotton. A smart engineer could study humans performing the task, make a list of the movements that were involved, and then try to create a machine that performed the same operations. In the beginning, the machine could be powered by a human, a horse, or perhaps a windmill, and then later on a steam engine.

The basic idea is that if a task can be formalized it can also be automated. This does not imply that all tasks will be automated. The formalization and the cost of developing the programs may be prohibitive. The solution is then often to automate parts of a job, keeping humans in the loop to perform the tasks that are difficult to formalize. Word processing is a good example. The user performs the advanced task of getting the syntax and semantics right—that is, the logic and contents of the document—while the computer works on the lower level of storing characters and aiding with the layout of the document.

It is often cheaper to employ humans to do the job than to invest in expensive machines. For example, we see that the degree of mechanization in agriculture is lower in countries where it is inexpensive to hire workers. Humans also have the advantage of being more flexible than machines, but the competition from information technology is increasing. Equipment and software are becoming cheaper and there are quite different ways to do things in a digital world. More than fifty years ago, very skilled workers were required to produce a high-quality car; today, each component can be manufactured within very small tolerances by using robots and other computer-controlled machines. High quality becomes the norm, even for inexpensive cars.

Tasks that are Ready for a Computer Takeover

Some tasks are easy to run on a computer. As we saw from the banking example, many tasks were formalized even before the advent of computer technology. The idea is then to convert the manual procedures into a programming language. In addition, the computer programmers must define how data is to be entered, stored, and presented.

Among the many applications that fall into this category are stock-keeping, hotel reservations, calculating insurance premiums, order handling, and accounting. All of these operations were performed under strict procedures in the pre-computer area as well. By converting these procedures into computer language, the tasks were automated. This was a major breakthrough. Before the first computers, an increase in volume, such as in the number of bank accounts, would have demanded an increase in staff. In old pictures of banks, insurance companies, and government offices we see hundreds of employees performing similar tasks. This is not the situation today. Today, computers can handle volume nearly for free. "Many" comes cheap with a computer.

Similarly, with robotics and advanced machines, fewer and fewer workers can produce more and more goods. Most of the tasks performed in a factory could be formalized and then automated. This has resulted in large layoffs in manufacturing. There is also competition from low-cost countries. But "low cost" may not be the right term; salaries may be lower, but the cost of machines and raw materials is similar all over the world. When automation is becoming the norm, offering inexpensive high-quality products, salaries will become a smaller part of the cost of manufacturing. The end result is that many companies today are in-sourcing, moving manufacturing operations back to their home country. While it may still be possible to produce the goods cheaper in a country with lower wages, other factors, not least the flexibility to react to market demand, will favor production closer to the market.[1]

In the early days of the computer—the 1950s, 1960s, and 1970s—the idea was largely to automate previously manual operations. In most cases, connections to the outside, to the customers, remained as they had been before, using letters or telephone, and the computers were hidden inside the organizations as stand-alone operations. This changed with the advent of computer networks, particularly the Internet. Now customers could connect and interact directly with the systems, the booking systems, net commerce systems, and so on. Since all these operations are formalized, they can be automated, as long as the customers have a user interface that they can master.

The Internet, with a browser that could connect to all servers using a standard language (HTML, Hypertext Markup Language[2]) and a standard protocol (HTTP, Hypertext Transfer Protocol[3]) did the trick. The rapid advance of these systems tells us that these application areas were ripe for automation; one only had to wait until the technology offered the necessary networks, user interfaces, and equipment.

Tasks that Need Additional Formalization

In principle, a task such as a hotel booking was formalized a long time before the advent of computers. However, the idea of a computer system is not just to replace the simple systems one previously had on paper, but to offer something much better. While I shall cover these issues in detail later, I offer one example here.

Back in the day, a hotel booking was performed by calling the hotel, the booking office of the hotel chain, or a travel agency. The booking office could use index cards or large charts to get an overview of available rooms. It was often a challenge to keep everything in order, especially if there were cancellations. With many hotels, many rooms, many reservations, and changes in reservations, errors could easily occur, leading to underbooking or overbooking. The advent of computerized booking systems, the first of which was made by American Airlines in 1952 (the SABRE system), made everything easier. As computers became faster, larger, and more reliable, they could more easily handle the booking process and maintain a good overview, independent of the number of rooms.

From the public's point of view, however, there was not much change. A customer still had to phone in to make a reservation. This changed dramatically with the Internet, which enabled customers to use their own PC, notebook, or smartphone, and access the booking system directly. Availability and prices were offered, as well as pictures, detailed descriptions, maps, tourist attractions, and more. The booking process is, of course, fully automated: all intermediates have been removed from the process. Today, the availability of good, simple-to-use interfaces that offer an immediate feedback is the norm. Most large hotels today are dependent on having an online presence.

We have also seen the rise of general third-party booking systems, such as hotels.com, booking.com, and airbnb.com. Their basic idea is that most customers look for accommodation based on location, and these sites all have the advantage that they can offer the customer a good choice of alternatives at nearly any location. The disadvantage of these sites is that the hotel has to pay a large fee for each booking. Most hotels and hotel chains also have their own booking systems; some have also opted not to be on the third-party sites.

System development may be expensive, but its advantage is the lure of automation, which leads to great savings in the future. Many companies have broken their back developing systems that have run over budget and over time. But there is little alternative: a hotel or a hotel chain can either establish its own booking systems, pay the general sites to do the job for them, or do both. A small number of hotels—the sort where the hotel itself is the attraction—may rely on booking by phone or email, but this becomes increasingly difficult when many customers are expected to get an immediate overview of availability and prices.

The advantage of developing one's own systems is that they may give a strategic competitive advantage—alas, it seems, only for a short time. When American Airlines developed its booking system, competitors followed close behind. The first banks to offer an Internet solution to their customers had an advantage, but most banks were offering similar solutions within just a few months. Today, large banks are competing to be the first to offer mobile pay systems, but they seem to have reached the finishing line together. Technology that is available to one is also available for the others. In many cases there is also a requirement for standardization, as with payment systems.

Interestingly, it may be easier for small firms to reach ahead using information technology than large firms. With fewer employees most tasks, from training to implementing the system in the whole company, becomes simpler. I have been engaged in developing software for several small and medium-sized companies. This has included full administrative systems, registering orders, planning, documentation and invoicing. The idea has been to use their own software for the niche parts—that is, for everything connected to production—and to use off-the-shelf systems for the standard parts, such as accounting.

For a foundry making propeller blades for ships, an advanced planning system, partly developed as a research project, managed to increase production by 20 percent. Just by running a smart planning algorithm, the company could utilize resources in a better way than before. Producing full documentation with the click of a button saved many hours of work each day. The main benefit, however, was to use information technology to change processes. For example, propeller blades are cast in a sand-fixture solution. A model, usually cut out of wood, is set into the sand to make a mold. Nickel-aluminum (bronze) is then poured into the mold at 1200 degrees Celsius. To counter the problem of the metal shrinking as it cools off, the model is blown up by approximately 5 percent. This handles shrinkage, but leaves too much metal on parts of the blade, which has to be removed through a cumbersome grinding process. However, with a set of smart algorithms we were able to estimate the shrinkage, thus producing ideal models that did not leave excess material. The total savings of this "shrink-to-fit" system were close to 50 percent. There is nothing to prevent the competition from doing the same thing, but while the competition may be assumed to be experts in the foundry business, they may not have the necessary IT competence. Thus, our company gets a real strategic advantage by employing systems that are not available to their competitors.[4] Large companies may not get this advantage, as their competitors (other banks, other airlines, and the like) will have the resources necessary to obtain their own systems.

While many companies have considered system development to be too expensive, a large set of good tools are currently available. These can take the form of flexible systems that can be customized to the given task, with easy-to-use development tools and premade modules that can be embedded in the programs. Thus, developing one's own systems may be feasible for many companies.

Some development projects will fail. Both big and small companies have spent large amounts of money on software that is never used. It is important to be quite clear about the motives and goals for developing the new system, gain an overview of all stakeholders, and describe how the new system shall interface with existing systems and processes. Experience shows that every dollar and man-hour used up front to analyze and clarify these issues will help achieve a successful result.

The Creative Part

A painter standing before a white canvas with brushes and a palette has the freedom to express anything—a landscape, a bowl of fruit, a portrait, and much more.[5]

The tools used by IT consultants, programmers, or system developers offer some of the same freedom as those of painters. Instead of replicating the solutions of yesterday, they can introduce more disruptive systems. Instead of improving efficiency by a few percentage points, they can offer radical new processes and perhaps solutions that can cut costs by half or more, handle greater complexity, or offer new products.

To achieve these results, it is important to listen to customers and work hard to understand their real problems. This is difficult. Customers are often strongly entrenched in their existing processes and may suggest solutions based on their limited understanding of what is possible to achieve with IT, in many cases influenced by the systems that they employ today or that their competition use.

In his famous paper, entitled "IT doesn't matter," Nicolas Carr called IT a "commodity."[6] He showed that companies use the same tools and the same consultants. Thus, Carr argued, IT is like electricity—something that we need but does not offer any competitive advantage. If we offer a standard solution to our customers, using off-the-shelf software products, Carr is correct. However, if we are able to think creatively, we may offer something else—that is, use IT to offer radical new solutions.

This is not always easy. There may be constraints that limit the freedom we have as consultants or developers. But there are many situations in which we are allowed to think creatively. If we use these opportunities, we not only provide value to the customers, but the development job becomes very interesting, perhaps to the degree where IT professionals feel as creative as (other) artists.

The opportunity to offer smart solutions may come in any type of application, as long as the following conditions are present:

- A good understanding of the customers' real problems
- A good understanding of methods and tools of the IT profession
- A willingness to think in new ways

As we have seen, the opportunity to devise creative solutions is not dependent on the size of a company. In many ways, smaller companies often have better opportunities for using IT technology than their larger competitors. Software development can be simplified if all users have the same platform—that is, the same type of equipment with the same operating systems. This may not be very easy to demand in a company with 50,000 users, but is no problem if there are ten or twenty employees. Small firms can also have simpler security and backup requirements than large firms. Further, they may use development tools that simplify development; an example is a complete package such as Microsoft Access, which provides a user interface tool, a database, and a programming language.

What is important is that the developers, the IT experts, and the users can work together without restrictions to define how processes can be achieved in a digital world. Sometimes the idea is just to automate existing processes, but there are often opportunities to radically change the way the job is done. The example from the foundry showed that savings of 50 percent were achieved in one of the most expensive and time-consuming tasks by realizing that computer technology opened up another way of doing a traditional process. This way of working requires an open mind and a willingness to avoid using the "we have always done it this way" argument. However, my own experience is that when one can offer new and more radical alternatives, customers can easily recognize the advantages and provide the go-ahead. The possibility of getting something better than the competition is always attractive.

Off-the-shelf Applications

While program development may be expensive and sometimes have an element of risk, it is always easier to use off-the-shelf products. These are systems that may be of interest to large user groups; examples include word processing systems, email, spreadsheets, and systems for handling photos. These are programs that are used by many, which implies that we get a lot of advanced software for a very reasonable fee.

When user groups are more limited, the licensing costs may be much higher. For example, the above mentioned foundry uses a program that can simulate how the metal will flow in a mold. The cost of licensing this program may be as high as the cost of development of other programs. Still, it will nearly always be cheaper to use an off-the-shelf product for standardized tasks than to try to develop a system. The disadvantage of using off-the-shelf systems is that the systems themselves and their benefits are also available for the competition. Again, it is like electricity: it is very convenient to have but it does not give any strategic advantages.

A great advantage of off-the-shelf systems is that they are used by many. This implies that errors in the software will be detected, and most often also fixed, immediately—perhaps even before you install the system. For popular systems there will be a whole industry that offers handbooks and user training. Some products have so many users that they become a de facto standard. For example, Microsoft Word is used in most companies, which means it is usually possible to change jobs without having to learn a new word processing system. There is also the advantage of being able to interchange documents in Word formats with others.

While the standard model for off-the-shelf software has been to buy the product, this is often changed to a subscription model today, where the software is installed from the Internet or used online. This offers continuous updates and simplifies maintenance.

Identification

A grocer will have a description of each product in a database, an airline will have a description of each ticket sold on a flight, a bank will have a description of each customer, and a hospital will have a record for each patient. Products, passengers, customers, and patients are most often identified by a code. Some of these may be particular to the computer systems, but many are standardized. Industrial products are given such a code—an EAN[7]—that identifies the country of origin, the producer, and the product. Similarly, books have also an international code system: ISBN. People may be identified by a national social security number or a personal number. The idea is to give everything a unique identifier. We should note that while products have an international code, the schemes for numbering persons are national.

The problem, then, is how to retrieve the code for the product, the passenger, customer and patient. That is, the person at the cash register will need to tell the system that the customer has bought this toothpaste, these apples, or this bottle of beer. The hospital needs to identify the patient, and the system has to identify a user when he or she logs on to a website.

This process is simple for products. It is done by writing the EAN code on the product in a computer readable form. A simple technique is to use a bar code, which can be represented as a font in a computer system, along with Arial or Times New Roman. Its advantage is that it is easy to make robust scanners that can read bar codes. Scanning the product at the cash register will then tell the system the EAN code of the product and information, such as price, may be retrieved from the database.

Identifying people is not as easy. While a hospital may give patients an armband with a bar code, this is not always possible in everyday life. Therefore, we have to use what we carry around with us: ID cards, credit cards, account numbers, or mobile phones. While these means of electronic identification are easy to use, their disadvantage is that we cannot be sure if they are being carried by the right person. In many cases, such as when boarding an airplane, we either accept that the card or phone identifies the correct person, or we require additional ID, such as a driver's license or a passport.

The electronic ID may also be augmented by secret passwords or a personal identification code (PIN), with the idea that only the rightful owner of the electronic ID knows the code. This is the standard method used for payments. However, we are now seeing PINs being done away with for smaller amounts. We shall discuss these systems in detail in Chapter 9. Passwords have to be remembered, so users tend to simplify these, for example, by using the name of pets, birthdates, and so on, which increases the risk that they could be guessed by someone else.

Another option for identifying persons is to use biometric data. The most common are fingerprint, retina, and face authentication. Fingerprint scanners are now an integral part of many laptops or mobile phones. As with most other technologies, there are drawbacks. One problem is that the fingerprint distorts as the finger is pressed against the reader. This impairs the system's performance and can result in the need to perform multiple scans or by accepting a reduction of the accuracy of the system. However, with an integrated fingerprint scanner in a smartphone, the advantage is a combination of both the device and the correct fingerprint. These systems will probably replace passwords in many situations.

Integration

While the owner of the first telephone had no one to call, we are now connected to a world-wide network, making it possible to call nearly anyone. The "network effect" tells us that the advantages of an application increase with the number of users. Think of Facebook, Snapchat, LinkedIn, and all the other social media services, or email or text messages. The full advantage comes when everybody is on.

We see the same effect for other types of computer systems. While the first systems for accounting, order handling, stock-keeping, and so on were stand-alone, these are now integrated. The advantage is that the same data can be used by all systems. Data entry happens only once, often automatically. If invoices sent by suppliers are digital, all of the necessary information may be retrieved directly from them, such as organization and account numbers, the total amount, and more. Suppliers may get their data from their own databases, based on a digital order sent by the customer in the first place. This digital order may have been set up automatically by a manufacturing system that has performed its job by looking at an overall plan and determining the components that are needed for each product.

At a supermarket or in most other shops, by scanning the bar code of the purchased items, all of the necessary information about a product will be available for the cash register, based on which the customer can receive a detailed receipt. The data can also be used to count down the number of items remaining. These data may later be aggregated to show what kind of products is sold on particular days to particular customers. Such "big data" can be used for planning, for targeted marketing, and for complex discount schemes.

When boarding a plane, we identify ourselves at the gate by a credit card, frequent flyer card, or, for example, by our smartphone, often as a visual QR[8] code shown on the display. The system can then find the booking information in its databases. The advantage for the customer is that it is not necessary to bring a paper ticket. Further, an app with the ticket may offer additional information, updates of boarding status, gate, and seating. The advantage for the airline is that they have 100 percent updated and correct information on all passengers on all flights. Previously, with the paper-based system, data on every ticket had to be punched into the system or scanned by an optical character reader in order to obtain data for statistics.

Interestingly, airlines have still not used their full potential to communicate directly with their customers. For example, a good computer system should be able to rebook passengers following a cancelled flight and send the result to the passengers as text messages, perhaps even before the cancellation is announced at the gate. This would reduce stress and the need to stand in long queues. However, here and in many other cases we see that it often takes years before a new technology is used to its full potential.

Change of Representations

Many companies and jobs are connected directly to a technology. We mentioned Kodak earlier. Its business model and that of many others were connected to analog image technology. Customers bought photographic film, put it in analog cameras, took their pictures, and had the film developed and copied to paper. This value chain involved everything from large companies such as Kodak and Fuji that produced film; companies such as Nikon, Canon, and Kodak that made cameras; laboratories that developed film; and small photo shops that sold the film. The digital technology disrupted many of these businesses. Some of the camera producers have survived, but are now facing a second disruption as the smartphone is taking over photography for most users.

Another example, also mentioned previously, is the music industry. As long as music was recorded on physical devices, whether it was vinyl or tape, the record industry could maintain its business model whereby music was paid for per song or per album. Record shops were an integrated part of this model. The model was maintained when music was digitalized on a CD, but lost momentum when it could be distributed online. While the industry did all it could to maintain the old model, the pirates soon offered much better service, letting consumers download or stream everything they wanted. Free music was less important than the flexibility of access. In the end, the industry caved in. Today there are many sites that offer unlimited access for a reasonable subscription fee. There is a place for both musicians and perhaps also the producers in the new model of streaming, but perhaps less profitable than before, except for the record shops. Their whole business model was based on a physical representation of music.

While digital music depends on a player in order to be heard, books have the advantage that they come with their own "viewer": paper. While a book can be read in digital form on a smartphone or a viewer, such as Amazon's Kindle, the paper versions are still popular. In some countries, such as the United States, competition from downloading of digital books and online bookstores has forced some physical bookstores to close. However, the picture is not as clear here as it is with music, as there still are advantages to using the paper version.

This may change with the advent of better viewers. The eInk version of the Kindle is interesting because it uses ambient light. The page is generated by moving electrically charged pigments on the display. The units are grayscale and have a six-inch screen and Wi-Fi connection. It works fine for books with only text, but has problems when showing large tables, figures, and photos. If we want to get rid of paper altogether as a medium for presenting information, we will need large viewers, better resolution, full color, and so on. There are many prototypes and a lot of research in this area; still, the Kindle and similar viewers have been around for 10 years without any dramatic change.

It is not only the bookstores that are dependent on the representation. University libraries have been around for 3000 years and universities were often built around the library. Today this is changing. With scientific books and scientific papers available online, researchers can retrieve what they need without leaving their office. Even large repositories of books can be scanned and moved online. In practice, we see that the university library is also based on one form of representation. When all data is available digitally, such libraries may fall out of fashion in a similar way to record stores; 3000 years of history is no guarantee that they will be here tomorrow.

Difficult or Easy? A Case

In this example we can choose one of two tasks. The first is to develop an inventory control system for a home freezer. The second is to develop such a system for the freezer storage of a large grocery chain. Which task will be the easiest to develop?

At first it may seem that it would be much easier to handle the few items in the home freezer—perhaps fewer than fifty—than hundreds of thousands in the large freezer storage. But remember that "many" is no problem for a modern computer system. Even a small laptop, or for that matter a smartphone, will have enough storage capacity to handle millions of records.

Further, and this is the important part, the large number of items in the freezer storage have required a large set of formalized procedures. That is, even before the advent of computers, these strict procedures were necessary in order to retain oversight. For example, all items will have an ID. The ID will be recorded on the package with other important data, such as weight, dates, and producer. The ID will probably also be offered in computer-readable form such as a bar code. Packages will be standardized and may be stored on pallets. Insertions and withdrawals from storage are often based on a pallet as the smallest units—that is, one can operate with a number of pallets but never with a part of a pallet.

The storage area itself will be "formalized" by naming all locations; for example, by row and column. These may also be represented as bar codes on the shelf. When inserting an item in the storage, the bar code of the product and the shelf will be scanned and registered in the system. Alternatively, the system can maintain a register of all free locations and then assign a random location to each pallet. Similarly, there will be strict routines for taking goods out of storage. The superintendent will be able to get reports at any time on what is in storage, free locations, goods that are about to expire, and more.

If we move to the home freezer, we will see that there are no standard packages and probably no well-defined locations. A family with young children may have a problem maintaining strict procedures, risking that items could be removed without notifying the system. Without bar codes, everything would have to be registered manually in the system. Packaging of some items, such as a bucket of ice cream, would allow for the removal of some scoops, in practice requiring that the bucket was weighed before it was put back into the freezer.

While it is not impossible to use a computer to keep track of the inventory in the home freezer, the cost of running such a system, with regard to keeping track of insertions and withdrawals, would be very high. In practice, it will be easier to just go through the freezer to get an overview. An advantage of this manual process is that there are a limited number of items.

In summary, our first impressions of what is difficult or easy may not be correct. IT technology requires us to think in new ways.

Taking Advantage of Computer Technology

The simple approach to computer applications is to automate manual processes. Forty years ago, this could have meant offering computer terminals to intermediates, such as travel agencies. Terminals gave the agencies direct access into the various booking systems, first for airplane tickets and later on for other tickets, hotels, and so on—clearly a breakthrough. But while communication between the agency and the booking system became electronic, the customer still had to communicate with the agency in person or by phone.

The real change came when customers gained direct access, which became technically feasible with the advent of the Internet. Then a customer could use his or her own network-connected PC with a standard browser. Based on the standard layout language HTML and the standard network protocol HTTP, any browser on any machine could connect to any server on the net. Customers did not need special software to connect to a reservation system or to connect to a bank or an insurance company as long as all systems followed the HTML/HTTP standard.

This meant that customers could do the booking, independent of intermediates. In the beginning there was a lot of skepticism; would an ordinary customer be able to perform the task without making errors? But the early systems proved that most customers could handle the system with ease. Improved user interfaces helped, and there were still agencies for very complex tasks. The advantage was, in most cases, a better overview of alternatives, prices, and conditions. In addition, one saved the cost of having intermediates in the loop. Today we see that more and more customers are doing the work themselves, booking tickets and administering their bank accounts or insurance. In addition to the improved overview, we should expect that most of us use less time than before on these tasks. A lot of time is saved when there is no need to visit the bank in person.

Improved functionality reduces the need for manual handling. In the beginning, banking systems could handle simple transactions such as a payment or a transfer from one account to another; now they can also handle complicated loan processes. A fundamental requirement is that background data are available online for the system. For a mortgage, this may be information on the property as well as income and financial data for the applicant. The system may not be able to handle all types of loans, but may take all common cases. In practice, these are also the frequent cases. Thus, the bank only needs manual handling for the more complex loans. "Complex" will be a dynamic variable here. As the software improves, it can handle more and more, including tasks that are now considered complex. In addition to the advantages with regard to efficiency, automated loan handling processes offer an improved customer experience. It now becomes possible to get an answer in minutes instead of weeks.

Most companies have used the advent of computer technology to automate activities such as order handling, procurement, and payment. Orders are sent out electronically, often in a format that can be read directly by the supplier's order handling system. When the goods arrive, it is easy to check directly with the electronic order that everything has been received. This gives the background data for paying the electronic invoice from the vendor. All these systems make the company more efficient.

Some large auto manufacturers use an alternative system. They can manage without sending orders, registering received materials, and getting invoices; instead of handling the documents more efficiently, they eliminate them by letting suppliers have direct access to the auto manufacturers' planning system. For example, assume that a tire supplier sees that the factory is going to produce one hundred model X cars. They will then know that they have to deliver 400 tires of the type that model X uses. The auto manufacturer does not have to check that every tire is delivered. The fact that the 100 cars drove out of the factory is proof of delivery. Further, the supplier does not have to send an invoice. The tire manufacturer will be paid for the 400 tires according to predetermined price agreements. Thus, they have used computer technology to eliminate the paper, offering a very lean and flexible system.

It is not always easy to see the opportunities that IT offers due to being too entrenched in existing procedures. Consider the Norwegian State's travel regulations as an example. In order to simplify travel expense claims, to avoid having to present a receipt for every amount, a per diem system was introduced many years ago. Without such a system, reimbursing any amount, even a cup of coffee, would require a receipt. Since expenses may vary by country, and even by city, there is a special amount for each country, and also for some large cities. Then it was discovered that this could be a source of non-taxable income; for example, if one visited friends instead of staying at a hotel. Parts of the per diem were then made taxable, depending on the type of lodging. This caused many interesting situations. When I stayed at a friend's house while visiting the University of Pittsburgh, I had to pay tax on most of the per diem. I could have avoided the tax by staying at a hotel, even just by paying for a room and getting a receipt. When my friend visited me here in Molde, Norway, he also had to pay tax on the per diem. This required him to get a social security number and deliver a tax form at the end of the year. Since the amounts varied by country and also cities, he had to offer detailed data about when he arrived and when he left. In addition, the system had to compute and report the part that was taxable, which also required detailed information about the type of lodging.

The problem here is that the new digital system was built on regulations that were made for the old manual system. The per diem was transferred from the old manual system to the new digital system without any discussion. While the per diem was a smart move at the time when it was introduced, when receipts were often handwritten, it is not needed today. With nearly all payments performed by credit card, one can retrieve all information electronically from the credit card company, and paper receipts are no longer needed. Reimbursement can be based on expense and the travel allowance form can be set up automatically. There will be no "profit" and therefore no need to include taxation.

This simple case illustrates a common and serious problem. Current processes are digitalized without realizing that the new technology offers the possibility for quite different solutions. This is one reason why many IT projects fail. People often fail to understand that the world, with all its processes, is a product of the tools that were previously available—pen and paper, typewriters, letters, phones, archives—and that new technology offers new possibilities.

In recent years I have been involved with creating an app for plumbers. Being a plumber used to be a practical job, installing washing machines, water heaters, and toilets, and connecting these to water pipes and sewage. Today, a plumber needs to document what is done and to inform customers, head office, and authorities. Some jobs require following safety procedures and meeting environmental requirements.

Many plumbers consider this to be annoying bureaucracy, especially as report-writing skills would not have been a factor in their choice of profession.

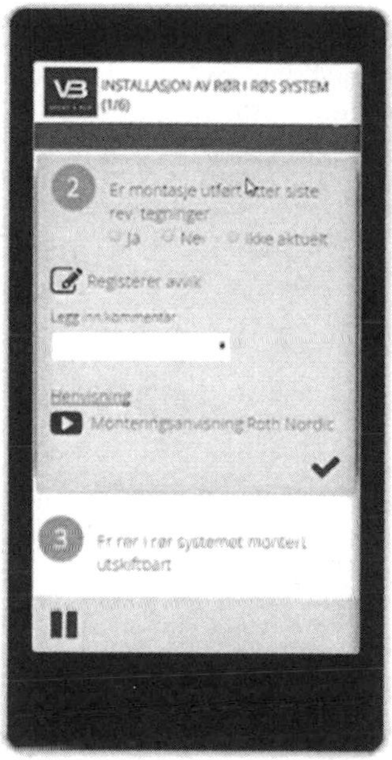

Figure 2.1 An app for plumbers (Norwegian version)

The app contains a *process list* for each task (Figure 2.1). This takes the plumber through the current job, which in this case is the installation of a pipe-in-pipe system. It acts both as a checklist and as a documenter. Data entry is simple: the plumber answers the questions, often just by choosing a yes ("ja") or no ("nei") option, writing some text, or choosing a premade text or taking a picture. When the job is finished, a report is sent to the head office and to the customer. The system will also communicate with the customer through text messages. These can be automatic, such as "starting the job" or "finished and leaving the premises," or manual—for example, if the plumber needs additional information from the customer.

This process removes a lot of the bureaucracy. At the same time, the plumber can be assured that all regulations have been followed. If there is a change in these regulations, it will be up to the person who creates the process list to make sure that these are updated. The plumber just has to follow the list for each type of job and is then guaranteed that all regulations have been followed.

Smart Smartphone Applications

With a smartphone in your pocket, you have access to all kinds of information. On the way to the airport you can check whether the plane is on time; upon arrival in a new city you can get directions to the hotel; and everything you need to know about an upcoming meeting will be available. This is all theoretical, of course. In practice, it takes a lot of input, pushing small buttons on limited keyboards, waiting for downloads, searching for information, and so on. We do not always have the time to retrieve the information (for example, when we are running to catch a train), or the opportunities to provide input may be limited (when driving, for example). In many cases, the effort required to extract information from the smartphone is so great that it becomes impractical, even with modern user interfaces. This is especially the case when we cannot devote all our attention to the device. Speech recognition systems may be of help in some cases, but these also have limitations.

Much of the research on user interfaces has concentrated on offering easy-to-understand interfaces. These may come in the form of menu systems, forms with fields for input and command buttons, or wizards that support users through a process. These research efforts have recognized that it is important to limit the necessary input, for example, by retaining important information from previous encounters with the user, since offering input from a keyboard, mouse, or touch-sensitive screen requires users to provide the input and correct the resulting mistakes. This takes time, and even if the cost (counted as the number of input operations) is low for one operation, it may be quite high if one considers all the operations that a user performs throughout a day. Therefore, it is important to limit the number of "clicks" where possible.

Amazon presents its one-click ordering system as the ultimate example of a simple-to-use interface. This is achieved by storing a customer's address and payment data. By implementing and marketing this scheme, Amazon has recognized the cost of clicks and that busy users find it important to limit input. The next step is more difficult, however. Can we offer zero-click systems? Clearly, a zero-click book order is not possible, as a consumer will always want to have a choice of what product to buy. But there are many other situations in which input-free user interfaces will be achievable.

In fact, such interfaces are actually quite common. A good example is a wristwatch, which tells us the time just by looking—no input is required. Similarly, signs offer information without requiring input. While we stand at a platform waiting for a train, a display may tell us when the next train will arrive and its destination; this information is important to the traveler and can be received without any input. Another example is modern smartphones that display the name of the caller. And, of course, we operate in physical environments where data is gathered just by seeing and hearing, such as looking out the window to determine the weather.

One can also explore new interfaces with a smartphone, such as a personal assistant that can push information that is relevant to the user and can initiate actions that it considers necessary.[9] An example of the latter could be to check whether the user will be able to catch a flight, using time data for traveling to the airport, time to go through security controls, and actual times for departure. If not, it could warn the user that a change of booking is required, and also offer to do this. The idea is to implement an assistant that can offer relevant information when it is needed, directly on the display of the smartphone.

The idea of assistants is not new—Microsoft introduced the widely disliked Office Assistant in the late 1990s and, as early as 1987, Apple CEO John Scully described the "Knowledge Navigator."[10] However, I feel that it is now practically possible to implement this idea. Both Google with Google+ and Microsoft with Corsera are developing such systems.

Smartphones have made such a development possible, and since most important data is available in digital form, it is possible to deliver relevant information. The job of the push system or "assistant" will be to select information from a repository based on a set of selectors:

- Time
- Location (based on GPS or mobile phone triangulation)
- Usage patterns

These selectors will have the greatest advantage when they are used together. For example, assume that it is 8 a.m. and the user is at home but has a scheduled meeting at work at 10 a.m. The system knows about the meeting (from the calendar), knows the user's current location (GPS, determined to be "at home"), knows the location of "work," and, from previous patterns, knows that it takes 45–60 minutes to commute between "home" and "work" at this time of the day. Since the user has asked for a fifteen-minute warning about events, information on the meeting will be offered to the user at 8:45 a.m. If the user had been at work, the information would have been given at 9:45.

With all data on the user available, and with background data on the Web, it will be possible to select the information that the user needs in many situations. On the way to the bus stop, the system can present the time that the next bus will leave. If the user has a hotel reservation in a city, the task will be to direct the user to the hotel. Users may have shopping lists for various stores, and when the smartphone detects that the user is close to one of these, the shopping list can be displayed.

Systems like these may give useful information with little or no input. The challenge is to get access to all data in a form that allows it to be used directly. Just like a personal assistant, the computerized version needs a complete overview in order to be able to give good advice.

Conclusion

In December 2010 IEEE Spectrum published a report with the "Top 11 Technologies of the Decade." Six of these were IT: smartphones, social networking, voice over IP, multicore CPUs, cloud computing, and digital photography; a further two—drone aircraft and planetary rovers—are largely based on IT. IT is now part of nearly every new technology, emphasizing that IT is a general technology that can be used in most areas. Even breakthroughs in other areas, such as gene technology,[11] are largely based on using smart algorithms and computers to process data.

In its first fifty years, IT invaded the work space; now it is also invading our private lives. Most people in developed countries own a smartphone. In Norway, for example, more than 85 percent of people over the age of 15 have a smartphone; for those under 50, the number is 95 percent.[12] With these numbers, companies, government offices and most others can expect that their customers will carry a computer connected to the Internet. With the infrastructure in place this opens the door for many new applications.

Many processes are simplified when data entry can be performed immediately. The smartphone makes this possible, as shown in the case above of the app for plumbers. For example, it can be important to document the state of the equipment and the pipes when starting the job. This can be performed by taking a photo. Since the app connects the picture to the current order and to the current point in the process list, such as the job start, no other input is necessary. The "assistants" that Google, Microsoft, and others are developing use a similar strategy.

Notes

[1]   The Spanish clothing and accessories retailer Zara provides a good example. Instead of outsourcing to low-cost countries in Asia, its factories are situated in Europe. Thus, Zara can claim that it needs just one week to develop a product and get it to stores, which is much better than the six-month industry average.

[2]   The first version of HTML was designed by Tim Berners-Lee at the European Research Community in CERN in 1989. With HTML one could describe the layout of a document, a web page. A central idea was that one could connect to other documents with a link, a URL (uniform resource locator).

[3]   HTTP defines the protocol for transmitting HTML pages. The main commands are the GET and POST commands. The GET method is used by the browser to retrieve a document, offering the URL as the location. The POST method requests that the server accept the document—for example, a message for a bulletin board, and store it under the web resource identified by the URL.

[4]   For a more detailed account, see Olsen, K.A. (2009). "In-house Programming Is Not Passé—Automating Originality," *IEEE Computer*, April.

[5]   Olsen, K. A. (2017) CreativITy, How Information Technology (IT) Can Be Used To Make Radical New Solutions For Customers, to appear in *IEEE Potentials*

[6]   https://hbr.org/2003/05/it-doesnt-matter

[7]   The EAN (European article number or international article number) is an international numbering system for industrial products. The number is often represented as a bar code on the product itself or on the packaging.

[8]   A two-dimensional bar code.

[9]   See Olsen, K. A, Malizia, A. (2011) "Automated Personal Assistants," *IEEE Computer*, November.

[10] Sculley, J., Byrne, J. A. (1987). *Odyssey: Pepsi to Apple*, HarperCollins.

[11] As a general book on the history of the gene, I recommend Siddhartha Mukherjee, *The Gene: An Intimate History*, Scribner, 2016.

[12] http://medienorge.uib.no/statistikk/medium/ikt/388

Chapter 3
Complex Computer Applications

Computers were originally used for mundane tasks such as accounting, interest calculations, and ballistic tables for the military. Perhaps inspired by science fiction, new and more radical applications were discussed even when the technology was very young. For example, a natural language translator between English and Russian was presented as early as 1954; it was a prototype, but the inventors expected to have a full version within three years.[1] In 1956 the field of artificial intelligence (AI) was born and in 1965 Nobel Laureate Herbert Simon said that within 25 years computers would be able to perform all the tasks that humans did in organizations.[2]

These and many other prophecies were overly optimistic. Although many of the world's best universities and researchers prioritized research into AI, few products emerged. Today the enthusiasm (or hype?) is back: we now talk about big data, machine learning and new AI systems.

However, one should expect that, with the dramatic improvement of computer technology in the last fifty years, we should already have an abundance of smart systems. The interesting questions are then:

- Why do we still mostly only talk about all these new applications?
- Why do we only have prototypes?
- Why are many of these products not available commercially?

In order to answer these questions we shall go deeper into two cases: natural language translation and autonomous cars. Both cases have long histories and both have received a lot of media attention today.

Before we go on, we should note that there are successes. Computers now play chess, Go, and Jeopardy better than human players. However, these games, like most others, operate in a formalized environment with strict rules and a clear goal. This is seldom the case in the real world.[3]

Natural Language Translation

Language translation is a difficult and expensive task. In principle, it should be possible to do this by machine. In its simplest form, both the input and output is text. To support the process, one has dictionaries and grammatical rules, as well as vast repositories of written material in many languages. Humans perform this task by having good knowledge of the two languages involved, and—importantly—by the fact that they get the meaning of the text. Thus, humans work on all levels, from the lexical (words), via the syntactic (grammar), to the semantic (meaning). The problem for the computer is that it does not "understand" the text. Therefore, in practice it is restricted to work on the lower levels, consisting of words and grammatical rules.

A semantic understanding of natural language is not easy to achieve. It is painfully clear that both dictionaries and grammar descriptions are far from complete. New words emerge on a daily basis and old words change their meaning. Even works of famous authors have sentence structures that violate grammatical rules. This means that natural language computer systems based on dictionaries and grammar structures can only take us part of the way. Many have argued that one needs to be a human being in order to understand natural language—that is, to grasp all the underlying context information that is so important in getting the right interpretation. "See you at lunch" may be clear to a coworker, while the computer will ask when, where, what, and why.

Before discussing translation, let us consider the much simpler problem of proofreading. There are several low-level tools available today. A simple spell checker can look up every word in a dictionary to find typos and misspellings. A grammar checker, even with current limitations, is an aid when writing in a foreign language. For example, the English grammar checker in Microsoft Word may find some of my is/are and has/have errors, but not all. It detects the error in "apples and oranges is used as dessert" but not in "apples and oranges, the latter imported from Spain, is used as dessert." And, of course, since these tools only work on a lexical and a simple syntactical level, they will not catch subtle syntactic errors or semantic errors. Thus, the sentence "They lived at London" stands uncorrected, as does "A forest has many threes."

To be able to find these errors we need smarter systems, systems that can detect semantic errors. If I write "he have a red car," the grammar control in my word processor will mark the error, noting that it should be "he has." If I write "he has a red far," everything is correct from a lexical and syntactic standpoint, even if the sentence is without meaning. While a human would point out the error right away, based on an understanding of what is said, the computer system will have a problem. The problem can be solved by giving the computer understanding. Although this is not easy, there is a circumspect method. The sentence can be compared to what other people have written. If we have a large text repository, then searching through it may reveal that many people have written that they have a "red car," and that no one has written that they have a "red far." Since car and far have similarities in spelling, our program may suggest replacing far with car. What we have done here is utilize "big data."[4]

This idea of following the majority is not new. On the way to a soccer match in London, I managed to get on the right train. From then on it was not necessary to ask for directions; I just followed the crowd. We also use this sort of "cumulative knowledge" when we follow trails, where the aggregate of many feet is manifested as a dent in the vegetation. In all these cases there may be individuals who diverge from the trail, do not follow the crowd to the stadium, or write differently, but these will not make a statistically significant impression.[5]

While it is difficult to use the wisdom of individuals, we may find an aggregate wisdom of all. This is what Brian Christian calls a "human paste" in his book on the Turing test.[6] This "paste" offers surprisingly good results in many situations. Bots— robot programs that act as if they were humans—use this technique to give the answer to a question as a human would have done. A common method for mimicking humans is to store previously encountered answers, all the input the bot has received from users, in a large database. By using statistics, they can often give (that is, "choose") surprisingly good answers. By maintaining the conversation with a user, the bot will get new feedback that will improve its repository of answers.

While these shortcuts to intelligent natural language processing can offer good results in many situations, it is important to be aware of the pitfalls. The idea of using what others have written does not really catch the semantics. Thus, bots work on a sentence-by-sentence basis, but do not "understand" the progress of the conversation. Brian Christian offers an example:

> User: Do you have a boyfriend?
> Bot: Not at the moment, but I hope to find one soon.
> User: I would like to be your boyfriend.
> Bot: That would be difficult, since I'm happily married.

In 2016 Twitter users managed to corrupt Microsoft's unveiled Tay, a bot that the company offered as an experiment in "conversational understanding." The idea was that the system should learn to engage people in "playful conversation." However, in just one day users had tweeted racist remarks to the system that then started to return similar remarks, often just mimicking the input, but it could also utter statements unprompted, such as referring to feminism as a cancer. We cannot blame the system—it utters these statements without any understanding.

Using majority vote for proofreading should be easier since the user has provided (at least a first version) of the sentence structure. While proofreading can include everything from correcting a few commas and spelling mistakes to including advice and corrections on subject knowledge and organization, a computer program can focus on spelling, grammar, and the correct use of words—that is, it can enhance the functionality of current spelling and grammar checkers.[7] Such an application will be especially advantageous for users who write in a foreign language.

Some colleagues and I have developed a prototype system that applies this technique and works on all languages.[8] The system starts by building a text repository. This is done by searching the web for text in the chosen language. Then every sentence that the user provides is compared to sentences in the repository. The system will find the "he has a red far" typo, also correcting "they lived at London" and "we had ice cream for desert." Factual information is also easy to correct, "Beethoven was born in 1870" will be corrected to 1770.

What all current proofreading systems, including the system described above, really do is to move the experience of writing to that of reading—that is, moving from recall to recognition.[9] Since most of us are better readers than writers, we may be able to determine which suggestions to follow. If the sentence provided by the user has a zero or low frequency in the repository, the system may come up with another choice for alternatives that occur more often, especially if the words have a similar spelling. For example, the program may suggest changing "I visited New Fork" (a ghost town in Wyoming, USA) to "I visited New York." Thus we must rely on the user to ignore these false indications—comparable to being on the train that takes supporters to a soccer game, but where we have a different destination. This is, of course, crucial to all language aids. We depend on the user to make the right choice.

Similar methods may be used for automatic natural language translation. Google's translator employs huge text repositories that are used to train neural networks that can find the best match.[10] The EU is an important source, as documents and reports are offered in many languages, where the translation has been performed by skilled human translators. With advanced techniques for searching, Google will compare the text that a user wants to what Google finds in the repository. Ideally, Google will find a similar text; if not, it may find phrases and other components that can be used. Even if the software is impressive, the end result is sometimes not great, but still good enough to get an idea of the content of a document in a foreign language. Today professional language translation is still performed by humans.

So what is the problem? While dictionaries and grammar rules exist, natural language is flexible. We may use words that are not in the dictionary or have a different meaning from the one offered in the dictionary. Also, the meaning of a word may depend on the meaning of the full text. A translator that only looks at the sentence, not the whole document, will often fail. Grammar rules are general and are often broken in the text we find in newspapers and books. The idea is that as long as the text is meaningful, it is accepted. Some authors even speculate by breaking rules—for example, by avoiding periods and commas. Further, languages are dynamic and changing all the time. New words emerge and old words change their meaning—that is, natural languages are not fully formalized.

As we have seen, work on automatic language translation started more than sixty years ago. Since then we have had a revolution in processor power, memory size, and storage capacity, and we now have the Internet. This has enabled Google to solve language translation by looking into their huge archives for previously manually translated text. Still, languages are so multi-faceted and dynamic that there will always be words and constructs that are not interpreted correctly.

Since there are an unlimited number of exceptions, one may wonder if there will ever be a fully automated translation system. Perhaps the not-so-good solutions that we have today are as far as we will get.

Autonomous Cars

There is a lot of interest in autonomous cars.[11] In 2012, Google founder Sergey Brin said that such vehicles would be commercially available by 2017; they were not. In Norway, the minister of transportation has said that there is no need to regulate Uber as we will have driverless taxis in a few years. This seems very promising.

The problem is that we have heard the "few years" at least for the last twenty years. Back in 1997, General Motors talked about having driverless cars on American roads within six years. The idea of a fully automated car goes as far back as 1939, when General Motors had a futuristic exhibit of a driverless car at the World's Trade Fair in New York. In the late 1970s and 1980s the emphasis was on fully automated highway systems. However, by the late 1980s increased attention was being given to the autonomous car.

While there are no commercially available autonomous cars that can handle ordinary roads, modern cars have a set of driver-assist features. These may warn a driver for not staying in a lane, or trying to change lanes on the freeway when there is a possibility of hitting another car or if the car in front slows down. In the latter case, some cars will also be able to brake if the driver does not. In the smarter cars, such as a modern Tesla or Volvo, the cars can also steer themselves if there are good road markers. We may expect that these systems will improve in the future—for example, making it impossible to pull out in front of an oncoming car.

Since there is always a licensed driver in the driver's seat, the driver-assist systems only have to solve the general case, leaving it to the driver to sort out any exceptions or difficult situations. However, there may be a problem with latency. The car may tell the human driver to take over, but several seconds may pass before the driver is aware of the situation and able to deal with it.

This is the situation today. Major car manufacturers and technology companies, not least Google,[12] are heavily involved in research, development, and testing of a fully autonomous car. We have examples of driverless trucks on company roads in Australia, cars that can park themselves, and of commercially available cars where the driver can relinquish control of the steering wheel given favorable conditions. While the prototypes are impressive and the driver-assist features are important, it is clear that autonomous cars on all ordinary roads are still a long way off.

The reason for this is that there are too many exceptions. The software in the cars may handle some of these, but not all. There is an analogy here to the effort to develop automated natural language translation systems. As we saw in the previous chapter, the systems are still a long way from replacing human translators. Autonomous cars will face similar problems, but the problems encountered here will be even harder. While an automated language translator can handle exceptions by not translating a word or by accepting a not-so-good translation, the autonomous car does not have an easy way out. The solution that we use with other types of machinery—to turn the power off if something goes wrong—is not an option here. Imagine an autonomous car traveling on a freeway in heavy snow. Sensors and cameras are clogged and the car cannot navigate. If it stopped, then within a few minutes it would be covered in snow and would face the risk of being hit by other cars.

Driving includes an element of calculated risk. Assume you have a five-hour drive in winter conditions. The road surface may be good, but there may be some ice on parts of the road. The prudent solution would be to go slowly, but then the five-hour drive would be extended and driving at half-speed would also cause problems for other motorists. Another example is a slow-moving tractor ahead that cuts the grass on the edge of a two-lane road. The tractor covers our lane, and since we are in a curve it is difficult to get a good view ahead. We have two options: we can overtake the slow-moving tractor or stay put. In the latter case a queue will form behind us, as we will make it very difficult for other cars to pass. We manage to get a partial view of the road ahead and decide to pass the tractor. Both of these cases—running at full speed on winter roads and passing the tractor—involve an element of risk. Would I take this risk as a driver? Certainly. Would I take it programming the autonomous car? Certainly not.

There are many such cases. We slow down if there are children next to the road, but perhaps not if we are passing adults. In many countries pedestrians have the right of way on marked crosswalks. As drivers we then have to analyze complex situations: are these people only walking along the pavement or is their intention to cross? A not-so-ideal solution is for the autonomous car to slow down for every crossing. A "formalization" is often required; in this case it could be to install traffic signals at the crossing.

Sheep and cows on the roadside may be an indication that we should go slowly, but not when you see the single strand of electric fence between the animals and the road. While it will be difficult for the autonomous car to distinguish between adults and children or to detect the electric fence, the problem can be solved by going slowly in all of these cases. However, that will reduce the efficiency of roads and make driving more troublesome for other drivers.

Some problems, such as hitting an animal, have to be resolved outside the car. These cases are difficult for an autonomous car to handle. What will it do when the sensors and cameras detect a collision and blood on the road? Will it call animal control to put the animal out of its misery, or will it call an ambulance to aid an injured person?

Detailed maps are a necessary infrastructure for autonomous cars. These will be a help for telling the car exactly where it is, and for letting the car obey all traffic rules. Detailed maps are not always easy to make, however, especially in less formalized environments. While traffic in cities and on major roads may be regulated by traffic lights and signs, this is not always the case for other types of roads. The "priority-to-the-right" rule, used in many countries to regulate road crossings without yield signs, is especially difficult to formalize. The idea is that if no signs say otherwise, we have to yield for traffic from the right. But this is only the case if the traffic comes from a road that is open for all—that is, we do not have to yield for cars coming from a parking lot, a private road, and so on. It is often difficult to determine whether the priority to the right rule is in effect or not. In my own home town there is a case where the experts disagree; some call it an exit from a parking lot, while others argue that it is an open road (they solved the problem by putting up yield signs). In practice, it may not be possible to formalize all of these situations and represent the results as a detailed map. Of course, human drivers face the same problem, but we are probably better at reading the intentions of the other driver, in many cases by using eye contact.

Autonomous cars need to know where they are, with much better accuracy than what GPS can provide. In addition to, or as an alternative to detailed maps, an approach is to let the car recognize its surroundings. Depth and distance data may be measured using Lidar, an optical laser system. The data can then be compared to previously archived data to determine the exact position. By using a learning system where the archive is continuously updated with data from other cars, one can also handle changes to the surroundings—for example, recognizing the construction of a new building. While this should work in most cases, it is possible to envisage situations where there are more dynamic changes to the surroundings—for example, work performed by large earth-moving machines or snow removal equipment working next to the road. In these cases, the autonomous car may not find the landmarks that it needs.

There are many such exceptions. Road work, heavy rain, flooding, leaves that cover road markings, accidents, and many other situations will strain the ability of the software in an autonomous car to resolve each case, especially as there may be types of exceptions that will not be known in advance. If the autonomous cars are to be used on ordinary roads, all of these problems have to be solved. The multi-headed troll in Norwegian fairy tales grows two new heads for every head that is cut off. Those who try to improve software in autonomous cars may find that they are fighting such a troll.

When a new technology emerges, the hype tells us that it will be used in all environments in a few years. What we often see is that the application will be limited to a few areas and that it will take several more years than expected before the technology moves from the lab to commercial applications. The case here is against autonomous cars on *all ordinary* roads. If we limit the application area to situations where we have good control of the operating environment, many opportunities may arise. We mentioned driverless trucks on company roads. We should also expect that autonomous vehicles could work in city centers where slower speeds and parking possibilities would simplify the operation. Freeway operation is also a possibility, at least on freeways that have broad shoulders, and perhaps also some surveillance where manual control becomes possible in exceptional situations. The operation of self-driving cars will clearly be easier if we formalize the environment, for example, by adding sensors to the road itself.

An example of a system that may work is autonomous buses. First, these use predetermined routes that may be "formalized" to a higher level—for example, by good road markers, signs on every crossing that determine who has the right of way, and traffic signals at pedestrian crossings. Special conditions and exceptions may be handled by running the buses under a central control system, where a human has the possibility to control a bus remotely. The idea here is to limit the exceptions and to have a good solution (human intervention) when the automation fails.

Some of these principles may be used also for personal autonomous cars. But it is one thing to control a limited set of buses from a central facility and quite another to use this solution for all vehicles. Unlike buses, private cars drive on all roads. Therefore, we will probably never see autonomous cars that can drive on any road under any conditions.

Big Data Analysis

Data is currently captured at the source, at the Internet site, the point-of-sale terminal, from sensors, and from many other sources. With huge and inexpensive data storage, these data can be kept forever. The question then becomes whether we can learn something from these data.

Department stores have long experience in collecting information from customers. By offering rebates they ask customers to identify themselves, for example, by loyalty cards. Thus, sales can be connected to individual customers. A customer who buys diapers would then be a target for other baby products; one who buys motor oil would be a target for auto accessories. That is, these data allow for more efficient marketing.

Banks can use your credit card information to analyze your finances. Based on this they may offer loans, other types of credit cards, savings programs, and much more. Airlines use bookings to analyze traffic patterns and this information will have value when planning new routes. Google can get a lot of data from the words that are used for searches. This information can be sold to other companies. For example, a company that has launched a new movie may be interested to know what kind of interest this has created, and Google searches are a simple way to get an answer.

Modern cars will have the ability to communicate with a central server. We can assume that, in the not-too-distant future, all cars will continuously send data on their position, speed, and destination. Today part of this information may be gathered by following the driver's mobile phones. These data may be used to infer road conditions. It may, for example, become apparent that cars avoid using a certain road. We will not be able to infer anything if this happens to just one car because the driver may have just changed his or her destination. However, after several cars have avoided this road, we may infer that the road is closed, at least temporarily and alternate routes can be offered. If cars at a later date start to use this road again, the system may assume that the road is no longer closed. Data from mobile phones also allow Google and others to present current traffic conditions on a map, using color coding to show the average speed on each road.

If we have similar two-way systems when walking in the mountains, we can put trails on the map just by registering where hikers go. The visualization of the trail, such as the thickness of the line representing the trail, can be a function of the number of hikers that have followed the trail using the same "function" as in nature where the hikers make an impression on the soil. A solitary hiker will not make any trail in nature or on the map, but many hikers following the same route will make a trail both in the physical and digital sense.

In the examples above, there is much to gain and little to lose. Even a false inference, such as marking a road as closed while it may be partially open, will not have severe consequences. However, using statistics to infer that we do not have to yield for cars coming from the right at a certain crossing may be much more difficult. That is, if the computer does not have the whole picture it may draw the wrong conclusions. Even after a hundred cars coming from the right have yielded, the next driver may insist on his or her right of way.

A partial view is problematic in many situations. A department store may know what you buy there, but will not have data on what you buy at other stores. If you buy diapers in store X, store Y may not know that you have a baby. You may go to Google to look at a hotel in Paris, France. From then on you may find advertisements for Parisian hotels whenever you are on the net. However, while the marketers have noted your interest, they will probably not have seen that you have already booked a hotel; from then on, the advertisements are only irritating. I was at King's Cross Station in London, so TripAdvisor sent me a message about eating places around the station. What it did not know was that I was off to Cambridge. A smarter algorithm could have detected that I had just arrived in Cambridge around lunchtime and used this to offer restaurant suggestions.

A website that gets the whole picture will have an advantage. If it can catch your interest at an early stage—for example, seeing that you mentioned to a friend on a social network that you are planning a trip to Paris—it can then target you with advertisements for airline deals. However, as soon as you have booked your flight, the advertisements will turn to hotels, and then to restaurants and activities once the hotel is booked. The drawback of such a system is that we may get this "big-brother-is-watching-you" feeling. Or perhaps we will just enjoy the fact that we are receiving good and relevant offers.

While there is a good possibility to learn from current data, it will be much more difficult to "mine" data collected over time. All data must be seen in context. Over time, this context will change for most types of data, which means it will not be straightforward to compare data from different time periods. The above example of the customer who is interested in going to Paris illustrates this problem. Once someone has hit the "book now" button, he or she may no longer be interested in any new offers.

Aggregates over many customers may be somewhat more accurate. However, these will also be context-dependent. For example, interest in going to a particular country may be affected by terrorist attacks, currency exchange rates, the risk of work conflicts, and so on. All this means it is not a straightforward computational task to use historic data to say something about the future.

Within a closed system, such as a company, it may be easier to control for context. However, even here we need to understand that the drop in production in October last year was due to maintenance, that our supplier had problems in June, and that the small increase in February last year was a consequence of the extra working day caused by a leap year. In practice, the information gathered from big data analysis will always have to be interpreted; it will be very risky to use this directly.

Machine Learning

We have seen how a computer needs to be explicitly programmed. Ultimately, the program will be translated into a set of zeroes and ones inside the hardware. This defines an unambiguous and exact definition of the computation. However, the process of getting to this point may vary. With traditional programming, the programmer will define the actions, but these actions will be dependent on the data—that is, the actual computation is a product of both the original program and the data.

Machine learning takes this another step forward. Here, the program itself can be modified based on the data, or in some cases, created directly from the data. Machine learning is used to find hidden patterns in data as a way of "teaching" a program to learn a task or to help the program interact with its environment.

Some years ago, I was part of a group that was asked to develop a system that could estimate the time of arrival (ETA) for a regular passenger boat service on the North West Coast of Norway. The idea was to use the position and speed of each boat to estimate arrival time at the different stops along the route. This is a very straightforward job that can be handled by a simple program. However, since the route also includes exposed areas where the relatively small craft may meet heavy seas, the vessels may be delayed due to bad weather. Instead of going thirty knots, the speed may be reduced to five. If you were aboard you could ask the captain when he expected to arrive. He could tell you that there is no reason to worry; they will go slowly only a few minutes more, until they have left the exposed area, then return to full speed.

Ideally, the program should be able to offer accurate data in these situations, just like the captain. This is achieved by sampling data on position and speed at regular intervals for each trip, creating a database that can be used to generate information for subsequent trips. That is, the program was able to collect "experience" in the same way as the captain. For example, the boat may go slowly in heavy seas when crossing an exposed area. Using speed, course, and position, the program could go into the database and find data of previous delays given a similar situation, thus computing a correct ETA.[13]

In the same way, an autonomous car could register all variables. If the human driver had to interfere in a situation, the program could learn from this experience and try to avoid the situation next time. The difficulty is in making a correct interference as the model becomes increasingly complex. Thus, the method of making simple automated inferences used in the ETA calculation described above may not be possible when there are many data sensors and many different types of data. For example, the human test pilot in the autonomous car may tell the car to slow down because he or she sees a group of kids playing with a ball on the pavement next to the road. The car will sense people on the pavement and may infer that it should slow down whenever this is the case. That is, the human sees important details that the car ignores. In these cases, one will need careful changes in the algorithm of the car, perhaps also improved analysis of the context, in order to be able to do the right thing.

A problem that is often discussed in the media is that of a runaway trolley that is on course to crash into a large group of people. The question is whether you are willing to push the man in front of you onto the tracks in order to stop the trolley, killing one but saving many. In the case of an autonomous car, this situation may be expressed as a choice between hitting a woman with a baby ahead or veering up on the pavement and hitting an elderly man. The situation may be interesting from a philosophical point of view, but not from a technical one. In these cases, the autonomous car would try to brake; it will not have the data to conduct any detailed analysis (identifying the mother with the child and the elderly man or estimating the consequences).

The Future Job Market—Only for Robots?

Since the Industrial Revolution, human workers have been increasingly replaced by technology. At the same time, society has been able to create new jobs. Engineers are needed to develop the new machines and skilled operators are needed for production, installation, and maintenance. Still, we have seen a dramatic reduction of workers in industry. Automation has made it possible to produce more products with fewer people. The surplus of workers has to go into other areas. Many end up in low-paid service work. A laid-off car worker may not be happy to get a job in a hamburger restaurant, perhaps with less than half of the pay he was used to.

Some of the savings of modern manufacturing can also be used indirectly to create new jobs in health care, schools, and universities. That is, a modern society does not need everybody to be employed in producing food and manufacturing products. Active taxation of the productive jobs and industries is often needed to finance this part of the economy.

Some people have predicted that 90 percent of all jobs will be taken over by computers and robots, while other reports estimate the number to be as low as 10 percent. A study from the University of Oxford set the number at 47 percent.[14] The number will be dependent of the successes in many new areas. For example, an autonomous car that can drive on any road will directly remove many jobs, including taxi drivers and truck drivers. It may also have a large impact on the car market.

Many customers may find that they can avoid having their own car when they can call one as needed. This also has the convenience of letting the car take us to our destination without the hassle of finding parking. While we may need some workers to load and offload trucks, parts of these operations may also be performed automatically.

However, as discussed above, in practice there is a high probability that autonomous cars will have so many limitations that they will not be able to drive on all roads. Still, we may get more subway trains that run without a driver; trucks may be electronically connected in large "trains" on the freeway, allowing one driver to control many trucks; and some bus routes may also be run by autonomous vehicles. While many jobs may be affected over time, the number may be closer to 10 percent in this area than 100 percent.

In industry, automation may be taken farther when one creates new factories. These may utilize all the advantages of computer and robot technology, without being constrained by existing buildings and equipment, which means the automation can go farther. The disadvantage is that building a modern high-capacity plant requires huge investments and therefore a large market. Existing industries often use a more evolutionary approach. Today, many of the quick wins for automation have already been collected. The remaining workers act as "glue" between the machines—for example by moving parts from one workstation to the next. There are opportunities to take the automation a few steps further. This is just a continuation of the evolutionary process and will not result in any dramatic reduction of jobs. Also, the next process to automate may require heavier investments than the simpler processes that were automated before.

An example is checking in at an airport. Previously, the operation of checking the ticket and putting tags on the baggage was performed manually at the check-in counter. Today, in many countries, most of these operations are done by the passengers themselves using machines that can scan the ticket or collect it based on an ID, a card, a mobile phone, or a booking number. The machine will print baggage tags that the passenger affixes to the baggage. The passenger may then drop off the baggage at an automatic counter, using a laser pen to read the bar code on the tag.

There are still some manual counters, which are mainly used for priority passengers, those who have full-price tickets or who are frequent flyers. Interestingly, manual handling is now seen as the luxury option. However, the manual counters are also there to handle exceptions. Some passengers may have problems using the automats due to impairments or lack of experience. Others may have problems with their tickets, may have lost a connection, or may have special baggage. In practice there will be so many types of exceptions that a counter staffed by an experienced person will always be the best option. While it is cost-effective to develop machines that handle standard cases, it may be prohibitively expensive to develop automats that can handle every case. As in industry, the quick wins are already taken out in a developed country and it is the harder cases that are left.

The formalization of the environment is also an important factor. While it is possible to have some degree of automation while building components for a house in a factory, this will be much more difficult when building a house on its site. Of course, automation may be increased here too if one uses pre-made components, but this limits the type of house one can build. Similarly, a robot may install fixtures in a modern business building, but a robot cannot come to your apartment to install a new water heater. The environment of most ordinary homes is not so formalized that one can use a robot.

This aspect of formalization is important in many other applications. The cashier at a grocery store may be removed if we let the customer do the job, but this requires a formalized system where every product is identified with a bar code. While an experienced cashier can handle an exception, such as an unreadable bar code, this may not be an option for a customer, which may cause delays and irritation for the next people in the line. A grocery store may also have items, such as fruit, that are not pre-packed; this requires customers to identify what they have bought. The purpose of letting the customer do the job themselves is to save money. In the end, the question of self-service counters will be a matter of expediency—in practice, of how formalized the environment will be. As in the airport case, there seems to be an understanding that receiving human service is an advantage, especially where an experienced human worker can give us efficient and correct service. That is, there may be hotels, restaurants, grocers, and department stores where most operations are automated, and there may be those that offer staffed counters as a service. Customers may be willing to pay a little extra to get personal service.

The last option is valid where the service worker is of value—that is, where he or she helps a customer to pay, check in, or register what has been bought at the grocery store. However, we do not usually need intermediaries when we do online banking or ticket booking. That is, full automation, not the type of automation where the tasks are left for the customer, will nearly always be the best option. For example, a hotel may offer automatic check-in by registering your smartphone as you enter the lobby. Instead of waiting in queue for human service, your phone will tell you your room number. It will also act as a key to open the door. In this case, the automation is performing the whole job; nothing is left to the customer.

Until now, the jobs that have been automated have been balanced out by the creation of other jobs. One can hope that this balance will be upheld. In many ways this is a question of how one wants to organize a society. From a broad view, there are always more jobs than people to fill them. A society that wants to offer a good education to all children, from kindergarten to university, that wants to provide universal health care, that will fund research and infrastructure, and more, will always be able to employ nearly everyone.

The Hype

Many large IT corporations engage in far-fetched development projects. The idea is perhaps less about developing commercial products than generating publicity. In addition, even the most hare-brained experiment (perhaps especially the hare-brained) may offer valuable insight. In many ways we should enjoy the fact that these companies can engage in speculation and play. The problem is how the media present these projects using headlines such as "In a few years, drones will deliver post, pizza, and books from Amazon," "In a few years you may call someone in Germany, you can speak English and they will hear German," or "In a few years Google will deliver broadband to the world using high-altitude balloons."

While it is often easy to develop a prototype and to conduct an experiment with IT, the difficult part can be solving all the problems one will meet when the experiment moves into the production phase and must meet real-world challenges. For example, is delivering packages by drones science fiction or reality? Amazon answers its own question by saying, "It looks like science fiction, but it's real. One day, seeing Prime Air vehicles will be as normal as seeing mail trucks on the road." One day is pretty vague. But let us look at the problems. First, only light-weight and small packages may be delivered in this way. We will still need trucks. The prototype system can only be used during daylight and cannot be used with heavy winds or in rain. This will limit the usefulness of the system. A customer may expect a speedy delivery by drone, but then a rain shower will delay the transport. What if the drones are under way when it starts to rain or become windy? Do customers need special "helipads" for the drones, and should the packages stay out in the open air until they are collected? Sure, there may be air space that is not used by planes or helicopters, but are the drones able to detect and avoid overhead lines? What happens if the drones encounter birds? Killing a threatened red-listed bird may be the last thing they do.

There are clearly many questions, from the legal to the practical, that must be sorted out before operation can begin. While it is quite simple to set up a prototype delivery—anyone with a drone can do that—there are major problems to overcome when the technology meets the real world. Amazon probably knows this, but technology is an interesting thing to play with, and there will always be a possibility that the "one day" may arrive. In the meantime, such ideas receive a lot of media attention. It is much more interesting to present drone deliveries as the future than to ask all the critical questions. However, when we look as far into the future as with this project, we may just as well predict that the future may offer a 3D printer that produces goods in your home. Thus, there seems to be limits to how far we can predict a technology change before it just becomes raw speculation.

Translating natural languages is difficult, as discussed above. Translating speech online is a more difficult task. It seems that we should learn to walk before we can run. While it is possible to show impressive demonstrations, and some not so impressive, it will be far into the future before these systems become practical in use. As with textual translation, this will not only imply that the systems get something right, but everything right.

Google's Project Loon may be the most realistic of these "science fiction" projects. The goal—providing Internet access to the half of the world's population that is still not connected—is admirable. It involves launching balloons into the stratosphere every thirty minutes. High-speed internet is transmitted to the balloons from the ground, relayed across the balloon network, and then back down to users on the ground. In the stratosphere the balloons have to survive in heavy winds, UV radiation, and dramatic temperature changes. This is a challenging problem. The advantage is that the competition for air space may be lower than for Amazon's drones.

Personally, I would be inclined to put autonomous cars that can work on all ordinary roads on this list of far-fetched projects—that is, the idea of letting these cars use ordinary roads in ordinary traffic. If we assume that the technology will be available soon, why do cities and countries still develop trains and metros that are driven manually? Would it not be so much simpler to develop an autonomous train than a car? Most metros and trains runs in protected areas and many of the difficulties of steering are handled by the rails. In addition, there will always be a system for central control.

Self-driving metros already exist in airports and in a few cities. Copenhagen has the M1 and M2 line that go from the north part of the city, through the city center to two destinations in the south, M1 to Vestamager and M2 to the airport. Control of the trains is handled by a central computer system. The trains are in no way autonomous, but they run without a driver.

The interesting question, then, is why aren't most metros and trains self-driving? While the examples we have in Copenhagen and other cities are proof of concept, there must be other reasons why this is not the norm. Even in the much simpler environment of metro lines and trains, there are still issues of security and economics compared to cars that may slow the development toward more automation. Perhaps this is also an example of learning to walk before running.

Conclusion

When the United States put a man on the moon in 1969, there was a lot of hype about manned space exploration to conquer new horizons, such as expeditions to Mars. If, at that time you had made a future prediction saying that manned flights would never go beyond the moon, you would have been seen as extremely pessimistic. Not every curve continues into the sky, literally in this case.

In the last 50 years, computers have revolutionized both the workplace and our private lives. Nearly every office table has a keyboard and a computer display. Most people in the developed world have a smartphone in their pocket. Even a simple job like cleaning the streets has been partly automated. At one time this was low-skilled work that anyone could do with a broom and a shovel. Today the work is done by a modern, computer-controlled sweeper with a skilled operator who controls the machine through a computer display.

While the hype tells us that autonomous cars will be here "in a few years," multi-story car-parks are being built all around the world—an investment that would not pay off if autonomous cars become the norm. There are always prophesies about what we will get "in a few years" and many of these are unrealistic. However, this book presents its own "in a few years" case, that of the cash-free society. I shall argue that this is a more realistic prophesy than many of the others.

Notes

[1] https://en.wikipedia.org/wiki/Georgetown%E2%80%93IBM_experiment

[2] Simon, Herbert (1965) *The Shape of Automation for Men and Management*, Harper & Row.

[3] For those interested in a more detailed discussion see Olsen, Kai A. (2013) *How Information Technology is Conquering the World*, Scarecrow Press.

[4] It is also possible to use Google searches for manual proofreading: Olsen, K.A., Williams, J. G. (2004). "Spelling and Grammar Checking Using the Web as a Text Repository," *Journal of the American Society for Information Science and Technology* (JASIST), vol. 55, no 11.

[5] Olsen, K.A., Malizia, A. (2010) "Following Virtual Trails," *IEEE Potentials*, 29(1).

[6] Christian, Brian (2011) *The Most Human Human: A Defence of Humanity in the Age of the Computer*, Penguin Viking.

[7] Harwood, N., Austin, L., Macaulay, R. (2009) "Proofreading in a UK University: Proofreaders' Beliefs, Practices, and Experiences," *Journal of Second Language Writing*, 18, 166–190.

[8] Olsen, K.A., Indredavik, B. (2011) "A Proofreading Tool Using Brute Force Techniques," *IEEE Potentials*, 30(4), 2011.

[9] Nielsen, J. and Molich, R. (1990) "Heuristic Evaluation of User Interfaces," Proc. ACM CHI'90 Conf. (Seattle, WA, April 1–5), 249–256.

[10] Translation platforms cannot replace humans (*The Economist*) https://www.economist.com/news/books-and-arts/21721357-they-are-still-astonishingly-useful-translation-platforms-cannot-replace-humans

[11] Parts of this chapter is from Olsen, K. A. (2017) "The Case against Autonomous Cars," to appear in *IEEE Potentials*.

[12] Google's parent company Alphabet owns Waymo, established in 2016 to work on autonomous cars development.

[13] Olsen, K.A., Fagerlie, E. (2011) "Adaptive Systems—A Case for Calculating Estimated Time of Arrival," *IEEE Potentials*, 30(2).

[14] Frey, C. B. and Osborne, M.A. (2013) The Future Of Employment: How Susceptible Are Jobs To Computerisation, http://www.oxfordmartin.ox.ac.uk/downloads/academic/The_Future_of_Employment.pdf

Part Two
The Cash-Free Society

This part discusses the various aspects of money and the technologies that are needed for making digital payments, and discusses the advantages and disadvantages. Norway, which like the other Scandinavian countries is close to being cash-free, is used as a case. What is happening in these countries today will be the norm in other countries in not so many years.

Chapter 4
Money

Money serves three purposes: it is a means of setting a value to objects, it is used to perform monetary transactions, and it is a way of storing wealth. Money can have many representations; for example, it may be represented in the form of precious metals, as coins, paper bills, or as bits in a computer. To be of use, money requires a fairly organized society. In this chapter we shall take a very brief tour of the history of money.

So why do we need money? For use in transactions, the answer must be specialization. One person cannot do everything. The first settlers in North America provided for many of their needs themselves. While it was theoretically possible to be independent of outside products (after all, the Native Americans managed this), it was more efficient to obtain some items, such as guns, tools, and some commodities, from the outside. This could be achieved by barter or by selling some of the home-made products for money and using the money to buy what was needed. Later on the same argument of efficiency could be used to specialize farther: some people grew corn, others caught fish, some made tools, others baked bread, became dentists or doctors. Now money was needed to have the economy running, both to have a standard system for setting prices and to facilitate transactions. The settlers had to stock up with commodities before the cold winter started. The abundance of products from the harvest was stored in barns, in barrels, and in cellars to be used during the winter. However, when more and more people lived close to a functioning market, as is the case today, this became unnecessary.

While most of us make do with a monthly salary or pension, there are sometimes situations when we need extra funds; for example, we may want to buy a house or a car. The funds can be built up by saving what is left of each month's income; if this is not sufficient, we can get a loan. This is another use of money: as a way of handling wealth. With money it is possible to transform wealth—represented as savings and loans—into, for example, a new house. If we have done a good deal, the wealth will be maintained, and perhaps also increased, in its new form. However, at any time, provided there is a market, the wealth may be turned back into money.

Before Money

The early nomadic hunters and gatherers had few valuables. Aristotle defined these as barter societies, but was probably wrong. The closed groups were limited in size and in many ways worked as large families. Perhaps "gift societies" is a better description. The hunters who had killed a deer did not barter with the rest of the group and instead gave away most of the meat. There was no other option. The form of hunting that was performed with simple weapons such as spears and bow and arrows needed cooperation from many. This was a cooperative in which every member performed a task. Some looked after the camp, took care of children, made clothes, or gathered berries and nuts. Sharing was the norm; there was no need for any complex transactions.

"Wealth" was stored in the objects themselves, as food, furs or firewood. There may have been some barter between groups, exchanging products for products.

This changed when people started to live in villages, in areas with more people. It is one thing to share a deer with a small group, but it is much more difficult to share with a large group, perhaps also with many strangers. Then the deer, livestock and other farm products could be used for barter. The disadvantage of barter is that the process is time-consuming. There is also a problem of valuation. How many sheep equate to one cow and how many sacks of grain equate to one sheep? In addition, barter requires some sort of symmetry. For example, you may have a sheep but need grain, and need to find a person who has grain and is interested in a sheep.

The Advantage of Money

Cities also offered an opportunity for specialization: there were butchers, carpenters, bakers, and many other professions. This necessitated a system for paying for services. A payment system was also needed for taxes and fines, to pay for administration of cities and countries, for the military and for the court. That is, one needed means to simplify *payments* and to access *wealth*.

With a single currency based on numbers, everything could be validated—that is, given a price. A sheep could cost a hundred units of currency, a cow a thousand, which would mean one cow is worth ten sheep. A loaf of bread may cost two and some apples one. With this method, everything could be calculated, from taxes to dowries. When the stakeholders agree on the validation, the prices, money in any form will simplify transactions. The person selling the cow can now take the thousand units of currency instead of ten sheep, then the thousand could be used for any purpose, from paying for bread to hiring a carpenter.

In practice, one does not need to have fixed prices. The person buying the cow may want to haggle and suggest a lower price than one thousand. However, with money as validation, one has a language for negotiating. Money opens the way for trading. A merchant may buy the cow for 900 and sell it for 1100 in another market. The basis is a common understanding of "value" defined as money. Money is now a yardstick that can be used to assign a value to all products.

Money is also convenient for storing value. However, there are many alternatives. Value may be stored as gold and silver, precious stones, as a herd of livestock, in land and property. The advantage of storing wealth as money is that it is readily available—that is, it has high liquidity. This will not normally be the case if your wealth is in the form of property.

The need for money in some form is clearly seen in situations where cash is not available. Then other forms of money emerge, such as cigarettes in prisons camps during the Second World War. The advantage of having some form of "currency" is so great that if one form is not available, people tend to find another.

Early Money

Early money took many forms, such as furs in the hunting societies of Northern Canada, or cowry shells on tropical islands. Salt has also been used as money. For practical reasons, money should be something that is easy to carry around and does not rot or disintegrate over time. This did not stop the inhabitants on the island of Yap from using large stone wheels for money, but this was perhaps more a way of storing value than performing a payment.

Money can be represented as something that has an intrinsic value, such as salt, furs, and to some extent gold, or only a symbolic value, such as paper money and bits in a computer. Independent of the representation, it should be reasonably scarce. If it is too abundant it will lose its value; if it is too scarce it will not be practical for payments and for storing value.

Gold, silver, and precious stones fulfill all of the above requirements. The value of gold is especially high due to its scarcity. The one problem with using gold or silver for payment may be the need to check the quality of the metal. However, with weights and some experience, gold and silver was a means of payment that was used over a long time. It was so ingrained that both metals later became the basis for paper money. Precious stones could also be used as money, but these were even more difficult to verify than gold or silver.

Both gold and silver have the advantage that they have an inherent value, when used in jewelry, rings, ornaments, gold leaf painting, tooth restoration, and inlays in products such as swords. Today gold is also used in electronic circuits. Another advantage of gold is that it is resistant to most acids and to corrosion. In addition, both gold and silver are easy to mold into coins.

Coins

The advantage of a coin was that it could be standardized. Its value could be engraved on the surface. The other side may have a picture or the symbol of the institution that produced the coin, often the king. With common trust in the value of a coin, these could be used efficiently for payments.

The first coins came into use more than 2500 years ago, independently in Anatolia (Turkey), Greece, India, and China. The most famous early coins came from Lydia (Turkey). They are 2700 years old and are made out of electrum, a silver-gold alloy. This was a breakthrough technology that changed the nature of doing business. Alexander the Great used minted coins to pay his troops across a large empire. The coins were considered sound money since it was known that they held a certain weight of silver. The Romans used gold and silver coins for hundreds of years. Around 1800 the Spanish had their gold doubloons, the English their gold guineas, and the French their Napoleons. The first US silver dollar was minted in 1792.

The value engraved on the coin was not always identical to the metal value of the coin; some measure of trust was also necessary in this economy. If citizens did not have trust in the value of the coins, the government could always force citizens to accept the currency. In practice, the economy is stimulated when the currency is accepted by everyone.

Most modern coins, at least those with higher denominations, have a metal value less than the denomination. Nickel and copper are often used. The cheap metals have also opened the way for counterfeit coins. For example, copies of the €0.50, €1 and €2 coins are all being produced illegally. Approximately 100,000 counterfeit euro coins are taken out of circulation each year. For the coins with the lowest denomination the cost of producing the coins may be higher than the denomination. The US one cent coin is an example. In 2016 the cost of producing one coin was 1.5 cents.[1]

Banknotes

While coins may have an intrinsic value, banknotes do not. However, at one time banknotes were backed by gold. That is, one could take the notes to the central bank and receive gold in return. As we shall see, the costs of maintaining this "gold standard" became too high. Today all currencies are what we call fiat money, which means they rely on trust.[2] The advantage of banknotes is that the bills are very cheap to produce, they can be made homogeneous, are durable, light weight, and easy to carry around.

China pioneered the use of banknotes more than 2000 years ago, first making them out of leather and later on paper. To avoid the practical problems of hauling large amounts of coins around, exaggerated by the fact that Chinese coins were made of iron and other heavy metals, proxy notes were first used as an intermediate form of currency. These notes could be redeemed for coins. Later, this paved the way for dynasties to print fiat paper money that was not backed by coins. The founder of the Yuan Dynasty, Kublai Khan, saw the opportunities and printed his "Chao" banknotes from around 1270.

Marco Polo, who traveled in China in the late thirteenth century, commented on the use of paper money:

> All these pieces of paper are issued with as much solemnity and authority as if they were of pure gold or silver ... with these pieces of paper, made as I have described, Kublai Khan causes all payments on his own account to be made; and he makes them to pass current universally over all his kingdoms and provinces and territories, and whithersoever his power and sovereignty extends ... and indeed everybody takes them readily, for wheresoever a person may go throughout the Great Khan's dominions he shall find these pieces of paper current, and shall be able to transact all sales and purchases of goods by means of them just as well as if they were coins of pure gold.[3]

The temptation to print more money than what the economy could sustain was difficult to suppress. The result was high inflation and loss of trust in the monetary system.

In Europe, the first attempt to issue banknotes was in 1661, when Stockholm's Banco issued a paper note. The idea was to be able to replace heavy copper coins with something more convenient, very similar to the situation in China four hundred years earlier. The allure of printing more bills than what the reserves allowed, forced the bank into bankruptcy a few years later.

Banknotes, in contrast to, say, gold coins with an intrinsic value, are subject to inflation. Since it costs next to nothing to produce a banknote, it can be tempting to print too many. This was a problem in China a thousand years ago. Scholars called the paper "empty money."[4] Strict discipline is required in order to limit the printing of money. Clearly, the Weimar Republic, Germany, after the First World War, and Zimbabwe recently have not had this discipline. They all experienced hyperinflation, a situation where prices rise from day to day and money has to be printed with larger and larger denominations. Note that the "storing of wealth" purpose of money will not be achieved if the inflation is high. It will also lose its efficiency for making payments. When German workers during these periods received their salary, they literally had to run to the store to use it before prices increased. Modern countries avoided this threat by having an independent central bank that can stand up to any wish from politicians to use more money than they have. Of course, money may be generated in ways other than printing banknotes.

Another drawback of paper money is that it can be counterfeited. The large difference between the cost of producing a banknote and the denomination motivates forgers. To make forgery more difficult, modern banknotes have complex engraving, watermarks, holograms, micro-printing, and a security strip of plastic or metal. Special paper is used, really made out of cotton, but forgers will bleach a bill of low denomination and reprint with a much higher denomination. Some countries, including Canada, Great Britain, and Australia, use polymer materials for banknotes. These are more durable and more difficult to counterfeit.

A common technique for avoiding inflation was to peg the currency to the value of gold—the gold standard. The unit in the currency was then based on a fixed quantity of gold. A citizen could then, in principle, go to the central bank and request that his or her notes be changed into gold. Since few people actually used this option, the value of all the banknotes could exceed the value of gold. In 1913 the US's central bank was required to have gold backing of 40 percent of its notes.

While a gold standard can check inflation, it has several drawbacks. Most important was that the reliance on gold limited the economy. While it did keep inflation low, it also kept everything else low and could act as a limit to economic growth. Many European countries abandoned the gold standard in 1931. The US suspended the gold standard in 1933, except for foreign exchange. After the Second World War, 730 delegates from all forty-four allied nations established the Bretton Woods system. The idea was to regulate currencies, in practice pegging them to the US dollar and gold. The hope was that these fixed exchange rates between countries would facilitate international trade. The system was brought to an end in 1971 when the United States terminated the connection between dollars and gold.

During the late 1970s and early 1980s, when many countries experienced high inflation rates, many wanted to reestablish the gold standard, but the central banks managed to control inflation by other means. For example, one may use contradictory monetary policy. This may be achieved by decreasing bond prices (such that citizens will change money for bonds), increasing interest rates and, for example, requiring banks to keep larger reserves. There will then be less money to spend, economic growth will be halted, and prices will not increase as much as before. However, there is a danger of overdoing this policy and causing the economy to go into a recession where gross domestic product is reduced, household incomes fall, and the unemployment rate rises. This has happened in Japan. In the mid-1980s the yen gained value compared to the dollar. This hit the export industry hard. A later bubble in the real estate market caused a financial crisis where stock values plummeted.

There are some advantages to having a floating exchange rate. For example, an oil-producing country may expect its currency to be devalued with regard to others if the price of oil is in decline. This will make exports cheaper and imports more expensive, thus stimulating its economy. Norway is an example. When the oil industry was heavily affected by lower oil prices a few years ago, the rest of the economy thrived. The "rebate" given to foreigners that pay in dollars or euros has stimulated the export industry to such an extent that many of the jobs lost in the oil industry have been replaced by other export industries.

As mentioned above, money needs to be represented as something that is scarce, but not too scarce. Gold fulfills this requirement. But there is a risk in pairing the availability of money directly to a metal, as this limits the amount of money to the gold that exists and that can be excavated in the future. The advantage of fiat money is that the central bank can adjust the availability according to what the economy needs. The "quantitative easing" that central banks have used to reduce the effects of a financial crisis would not have been possible with a more fixed amount of money.

The idea here is to get more money into the economy in order to stimulate growth. This is done by using the opposite policy to that used to reduce inflation. Instead of selling bonds, the central banks buy bonds, thus inserting more money into the economy. Interest rates are decreased rather than increased, offering incentives both for businesses and private individuals to take out loans. Again, the medicine has to be given in the correct dosage. If the dosage is too low, one may not get the necessary stimulation; if it is too high, inflation may increase.

Cash Is Inefficient in a Digital World

Cash can be very effective as long as it is the only system used for payments—that is, when salaries are paid in cash and cash is used for every transaction, from paying rent to paying the grocer. The grocer can use the cash that he or she receives from customers to pay his or her suppliers and employees. Cash flows through households, retailers, and companies. Surplus cash can be deposited into a bank account and retrieved when needed. The banknotes and coins are agile. Even if the processes are manual, the costs are low if cash is used in nearly every transaction.

However, this is not currently the situation in many countries. Salaries are paid into bank accounts. Invoices are paid by an account-to-account transfer. A retailer who is paid in cash will need to move the money to his or her bank account before employees and suppliers can be paid. If the customer pays digitally, this transfer is automatic and potentially instantaneous.

Cash and digital payments may coexist in an intermediate period, but in the long run the digital solutions will take over. As fewer and fewer customers use cash, it will be more difficult to obtain. Banks will close down unprofitable cash handling operations, initially by moving these to ATMs and later by abolishing cash handling altogether.

A hundred years ago we would have found both trucks and horse-drawn carriages on the roads. A picture from New York taken in 1900 shows streets full of horses, but with one car in the corner.[5] Thirteen years later, another picture shows streets full of cars, with one horse. That is, even when a new technology is a clear improvement over an old, the transition will take time—society will need to adjust to the new technology.

Today we see that email is replacing letters in the post. It took many years from 1969, when the first email was sent, to get to this point. Today we see that many users prefer texting over email, at least for shorter messages. This development may go faster as the technology is already in place. The transition from buying music on physical media to online streaming offers another example of a technology shift.

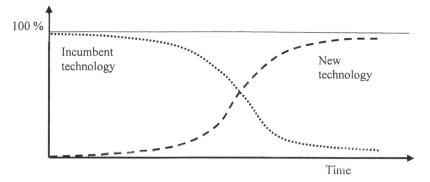

Figure 4.1. A new technology replaces an incumbent technology.

We often see a situation, as in Figure 4.1, when a new technology replaces an incumbent. In the beginning, the new technology comes in as an alternative, used by the early adopters and those customers who have a real need for something better. Then, as the advantages of the new technology become apparent to more and more users, it will take market share from the old technology. Of course, in these years the producers of the new technology will be able to put out better and better products. At some point, the advantages of the new technology become so clear that there is a very sharp change from the one to the other. If we return to the example of cars and horses, the changeover occurred at the point where cars became less expensive, more reliable, and customized for many applications. Roads were improved to make way for all the new cars and trucks, loading ramps were made suitable for trucks, and gas stations became common. Trucks now participate in long-haul transportation. At the same time, the drawbacks of the earlier technology, the horse-drawn carriages in our example, become apparent. They have less loading capacity, less speed, and cannot participate in many of the operations for which one uses trucks.

At this point, when nearly all users have moved to the new technology, the rate of change may be slower. There may be stakeholders who find it preferable to stay with the old technology, or there may be tasks for which the old technology is still competitive. We see this clearly with music: while 98 percent of the music played in most modern countries is streamed, cassette tapes, CDs, and even vinyl records are still in use.

Since all data in a monetary transaction, such as a payment in a store, is formalized and symbolic, the transaction can be performed digitally. The disadvantage of cash is that it is physical. In a way, cash payments hold up the whole system. While the cash register has the price that the customer has to pay in digital form, a customer paying with cash must convert this number into the correct amount or a higher amount, represented with physical notes and coins. Then the cashier must count the money, collect and return the necessary change, register that the full amount has been paid, and then store the money in the cash register. At the end of the day the contents of the cash register must be counted and checked according to the registered amount.

At one time it was argued that cash transactions were faster than electronic ones, but this is not the case today in most countries. The exception may be where the amount is small and the customer has the correct coin or banknote. Then cash is expedient. With broadband connections, the time to insert a card and type in the PIN is negligible, clearly faster than cash transactions that require change. With new technology, such as tap-to-pay using cards or mobile phones, a digital transaction will always be at least as fast as paying with cash.

In most countries, the cash that customers use for paying will reside passively in the cash register, apart from the portion that is used to give change to other customers. At the end of the day, or perhaps at intervals during the day, the cash will be counted, registered, and stored in a safe location until it can be moved to the bank. All of these operations must be secure. That implies that money is counted several times by different people, that withdrawals from the cash register are documented, and that money is stored in safes and transported in armored cars by security officers.

For department stores and most other receivers of cash, this has become a nuisance, a form of payment that generates a high overhead for the stores. A problem is that banks are losing money on handling cash. Customers are used to a system where cash handling is free of charges, so banks find it difficult to ask the customers to pay. If a person enters a bank to withdraw a thousand dollars from his or her account, the person expects to leave the bank with exactly this amount, without having to pay a fee. The solution for many banks is to reduce the service around cash. This has been done gradually by closing branches, removing cash handling from the remaining branches and, a temporary expedient, relying on ATMs. However, ATMs are also an expense for the banks and the temptation to remove these is also high.

Traditional banks are also competing with pure Internet banks that avoid all cash handling expenses altogether, and with large stores, such as Amazon, that offer their own bank-independent store cards. When more and more customers use digital payments it becomes easier for the banks to stop using cash. Fewer and fewer customers are affected by this. When we later look at Norway as a case, we shall see how far this situation has come in a country that is soon to become cash-free.

Conclusion

In this chapter we have looked at the advent of money and how various representations have been used through history, ending up with banknotes printed and controlled by a central bank as the most recent touch. In a modern society, money is mostly bits in a computer. The advantage is that it is then disconnected from any physical representation. It can be stored in a computer or transmitted over a network. Digital money is an integrated part of a digital society.

The transition from cash to digital payments will take time, as is the case with most technology shifts. In the beginning, the new technology may be offered as a prototype only and often needs further development to be of use. While early birds may be willing to fumble with any system as long as it is new, the larger user groups will demand easy-to-use interfaces. In many cases, speed is an important variable; processes performed in the new technology should go faster than in the old. Most importantly, the services should be available everywhere. A music lover who moves from buying CDs to streaming would want to play the music on the equipment of his or her choice: a smartphone, tablet, PC, at home, or in the car.

For applications that offer music, movies and, books, for example, a problem for start-ups will be the ability to offer a repertoire as good as that offered by the old technology. In these cases the intellectual property rights will be held by the incumbents. These are not usually the advocates for a shift in representation from physical to digital. Interestingly enough, this is also the case for digital payments. Later on we shall see that the national banks are not eager to move to digital technology.

When the new technology attracts larger user groups, we see that the infrastructure must be upgraded. While it is possible to send data traffic over a telephone line, this technology cannot offer the speed and quality that we demand today, for example, to stream video. Customers in modern countries will demand fiber cables to their home, local area networks in place to serve mobile equipment, and will expect that devices will be online at all times. When in place this general digital infrastructure will be available for both existing and new applications.

Rules and regulations will also have to change. At one time it was necessary to sign a check or a money transfer form; this has now been replaced by a secure login to the systems where these operations may be performed. For symbolic products, such as rights, subscription models are replacing the pay-for-each-item model. While most users seem willing to pay a lot for the equipment and the infrastructure, there is often an expectation that the systems should be free or cheap to access. However, this is now changing. The sites that offer music for streaming see that a large part of their customers are willing to pay for a premier service. We shall later see that no-fee models will be a good incentive for letting customers use digital payments for any amount.

As society opens the door for the new technology, the old may be left behind, as the cost of retaining both technologies may be too high. The market will usually take care of this. When customers stream music, record stores close down. When TV becomes high definition (HDTV), older formats disappear. When roads are made for cars and trucks, there is less room for horse-drawn carriages. Now, when payments are made digitally, there may be no place for cash.

Notes

[1] The republic of Transnistria, a self-proclaimed state that is officially a part of Moldova, has introduced plastic coins.
[2] From Latin: let it become, it will become.
[3] Marco Polo (1298) *The Travels of Marco Polo*, Penguin Classics, 1958.
[4] Ralph T. Foster (2011) *Fiat Paper Money, The History and Evolution of Our Currency*, Foster Publishing
[5] https://twitter.com/fastned_eu/status/584675134193164288

Chapter 5
Uncle Joe's Island

Many of us dream of inheriting a lot of money from a distant relative. In this chapter, I shall explore such a dream. One day you receive a letter in the mail. It is the will and testament of your old uncle Joe, the one that you always quarreled with at family reunions. He was an ardent old-fashioned capitalist and did not have any patience for your ideas about the digital society, especially not for digital payments. Uncle Joe wanted to have the money in his hands, and stacks of it in a large safe at home.

In Joe's will he has bequeathed you an island in the Caribbean. It is a small island, but it has a hotel and some other facilities and, according to Joe, a lot of problems. Since he has never been able to fix these, he now offers you the opportunity to try.

At least you have a good excuse to take some time off and go to the Caribbean. The closest airport is on the main island; from there you take a small passenger boat to the island. To your astonishment, the captain calls you "boss." Apparently the ferry is a part of your inheritance.

The island is beautiful but small and has a fairly large hotel. On the beach, not far from the hotel, are a small restaurant and a bar. In a meeting with the hotel manager you find out what their problems are. The currency on the island is US dollars and guests pay with cash while on the island. There is no Internet connection, so digital payments are not an option. On several occasions in recent years, criminals arrived by boat and robbed the bar and restaurant at the beach; on one occasion the hotel was also attacked. In addition, the hotel manager suspects that some of the employees may be skimming parts of the proceeds in the restaurant and bar. The manager has tried to overcome these problems by increasing security, but that is expensive.

While cash use could be limited at the bar and restaurant by letting hotel guests charge expenses to their room, this would not work for the high number of day visitors, who are tourists staying on the main island. They come with the ferry, stay the whole day at the beach, and return by night.

At first you think that this is an excellent place to test out your ideas about a digital economy, but the lack of a reliable Internet connection is a drawback. However, after exploring the problem you find that you do not really need the networks and the terminals. In fact, you can solve the problem with traditional means.

Your Own Currency

Together with a good designer and a printer, you develop a new currency—Island dollars. These are offered in the same denominations as the US dollars, but you skip the coins, using one dollar as the lowest denomination. Each note has a beautiful island scene. In addition, the notes have most of the security features of modern bank notes. To show that you are really modern, the paper is replaced by polymer plastics, which also enables guests to keep the notes in their bathing clothes. Your prerogative is to sign the notes. Since the notes are meant to be used at the island only, you are not afraid that customers will not trust the currency. This is a means of exchange only, not a way of storing value.

In your new system the ferry captain will change US dollars into Island dollars. This can be done either by offering US dollars in cash or by the use of a credit card. Most tourists use their credit cards. This offers no problem since the Internet connection is very good around the harbor on the main island. To keep things simple, you offer a one-to-one exchange rate.

The Island dollars can now be used on the island, replacing the US dollars. Since the Island dollars can be used on the island only, there is no incentive for robbers. Your control of the exchange process has also stopped skimming at the bar and restaurant. There is always a risk of forgery, but in practice you demand a receipt for transferring Island dollars back into US dollars. That is, the customer needs some sort of proof that this is a repeal of an earlier transaction.

Tourists can convert their Island dollars back into US dollars when they return home on the ferry. However, many keep a few notes as a souvenir. This is especially the case since the idea of Island dollars received a lot of media attention. The banknotes are something one can show people at home, and many want to have the complete series. Others don't bother to convert back and some want to keep them until next time. In fact, the advantage of offering your own currency in a series of beautiful banknotes is so great that you drop every idea of moving to a digital system.

Since the national currency in the country where your island is situated is blighted by a high inflation rate, dollars are often used as an alternative, but there are not that many dollars around. After a while you find that taxi drivers, bartenders, and a few shops in the mainland are accepting Island dollars from tourists who have forgotten to change back. You are also inclined to offer a part of salaries in Island dollars. While employees do not have the possibility of changing these into dollars, they can use them on the island and, to a limited extent, also on the mainland.

An advantage of Island dollars is that they are pegged to the US dollar, while the local currency may depreciate. However, you understand that the authorities may allow a small influx of Island dollars, but will not tolerate any alternative currency. Therefore, you are careful to only sell Island dollars to tourists who are on the way to the island. If not, you would also be back to the early robberies.

Seigniorage

This is your bonus. You established the new currency to eliminate robberies and skimming. Now you see that part of your income is from seigniorage. Each note a tourist takes home as a souvenir is a direct income to you. You sold the notes for their denomination value and were paid in US dollars. Each note that is used on the mainland is an interest-free loan for you. Of course, the day may come when the notes are presented at the island, but until then the loan that you received when you sold the note is interest-free. Your only costs are for printing the notes, which is marginal compared to their denomination—even for the smallest bills.

Your task is to balance the number of notes that you put into circulation. This is no problem for the currency conversion that takes place on the boat, as the Island dollars sold here are "backed" by US dollars. However, you have to make sure that the part of the salaries that you pay in Island dollars is limited in order to maintain the value of the currency. The employees accept this arrangement and you end up paying salaries in dollars and using your own currency for bonuses.

In many ways your Island dollars are similar to other currencies that are used in closed environments. Today these are usually virtual, such as the initial coin offerings (ICOs) that have become so popular today (we shall return to this in Chapter 11). But you see the advantages of keeping the currency physical. Even if broadband arrives at your island, you will be reluctant to give up the advantages of printing your own money. However, you will have lost your argument with Uncle Joe. We may wonder why many central banks want to maintain a cash-based economy in a digital world. Part of the answer is here—cash offers seigniorage.

Exchange Rate

An option that you may want to contemplate is changing the one-to-one exchange rate. Since you already control prices on the island, this does not make much sense, but by offering more Island dollars for one US dollar, you can reduce the value of the outstanding Island dollars. Tourists who return with Island dollars would then find that they are not worth as much as before. However, it will probably be smarter to retain the one-to-one exchange rate as this will maintain the trust in your currency.

However, let us assume that your Island dollars are used more and more on the mainland. This allows you to pay a larger part of salaries in Island dollars. Then, a change in exchange rate can be a way of increasing or lowering salaries with regard to US dollars, without changing the actual amounts that are paid. This is primitive psychology, but it works. Lowering the value of the domestic currency with regard to more international currencies will, in practice, reduce the buying power of a salary, especially for traveling abroad and buying foreign products. However, a lower buying power is a complicated issue, while an explicit reduction in salary is not. While we as employees accept the former, we should go on strike if the latter occurs.

For the case of the argument, assume that, in a period where it is difficult to attract customers, you set the exchange rate at 1.1, giving 1.1 Island dollars for every dollar. If you maintain prices, this will offer visitors a 10 percent discount. But it does not cost you as much to offer this rebate. The part of the salaries and expenses that you pay in Island dollars will be the same as it was before. Only the part that you pay in US dollars will be affected. In the long run, your employees and the suppliers that you pay in Island dollars may find that the value of the currency with regard to US dollars is lower than before, and may ask for a salary increase and higher prices. Hopefully, the tourist market will have improved by that time.

Countries that have their own currency will see similar effects. We mentioned Norway earlier. When the price of oil fell in 2014, the Norwegian krone became cheaper with regard to dollars, euros, and British pounds. This was a clear stimulant for Norwegian export industries and the tourist trade. Suddenly, all exports became cheaper and tourists stopped complaining about high Norwegian prices. The drawback was that imported goods became more expensive, but this is also an advantage for national producers. Switzerland is an opposite example. There, the value of the Swiss franc increased with regard to the euro, causing problems for the export industries.

Inflation

One year, a hurricane destroys part of the island. While buildings are insured and rebuilt, the tourist traffic drops dramatically and you are out of funds for a while. In order to be able to pay your employees and other expenses, you start printing Island dollars. This may work if you are very careful not to print too many. Otherwise, employees, suppliers, and others will find that they will have to pay more when using Island dollars. Your employees will demand to be paid in US dollars or a higher amount if they are paid in Island dollars. Again, this may require you to print even more money.

However, you know everything about inflation, and know exactly what happened in Germany after the First World War and the situation in Zimbabwe recently. Therefore, you maintain the volume of Island dollars at levels that you can control. In this respect you are performing the same tasks as central banks, to maintain the confidence in the local currency. You print more Island dollars when you are in need of funds and then, when the situation improves, you start to buy the Island dollars back. Of course, this may cost you a lot of US dollars, but just the offer to change Island dollars into US dollars (at an exchange rate set by you) will increase the confidence in the currency.

Conclusion

The idea of this chapter was to simplify the complicated topics of seigniorage, exchange rate, and inflation by putting you in charge, and limiting the scope to a small island. However, as we have seen, the same effects are apparent in large countries. We shall return to these issues in later chapters.

Chapter 6
From Analog to Digital

This chapter follows three historical and parallel developments. The first focuses on the development of alternate means of payments other than cash, such as checks and credit cards, as a run-up to the digital payment systems. In the second we shall look at standards, such as product codes and bar codes. Interestingly, these standards, which are so important for a digital society, came many years before the Internet and the www technologies that really made use of them. The third trend is the development of computer technology. When these three trends—credit cards, standardization, and technology—merge we have the complete foundation for a cash-free society.

In the analog world—that is, before the advent of digital computers—paper was the most important medium. Cash was on paper and the early replacements of cash were also paper based. Even the first credit cards relied on paper. While the card itself could be plastic, the imprint of the transaction was made on a paper slip. While paper has many advantages (for example, anything can be printed on a piece of paper), the disadvantage is that all further handling must be physical. Checks and credit card slips have to be transferred physically to the banks, where they must be counted physically. There were some possibilities of using machines in the processing, but these were limited by the fact that much of the data was in handwriting. Later on, as computer scanners and optical character recognition became more advanced, some of these processes could be automated.

In many countries, paper is now out of the payment loop. Checks are only used rarely and all credit card transactions are digital. However, it is still interesting to take a look at these "early technologies" as a means of explaining the development toward a digital society. We will also see that a set of standards are necessary when formalizing the payment systems. Then we have the background for using the development of computer technology to automate the complete payment system.

Checks

A check is a letter asking a bank to pay a specific amount to the bearer of the check, deducting the amount from the account owned by the person who wrote the check. Checks have been used for more than a thousand years. In fact, early versions of checks were in use by the Romans. The first preprinted check forms were issued by the Bank of England in 1717. The Bank of New York issued the first checks in America in 1784.

The use of checks peaked in the second half of the twentieth century. Instead of paying employees in cash, which had been the norm, banks persuaded companies that it was simpler to pay using checks. A clear advantage was that one no longer had to operate with large amounts in cash, transporting cash from the bank to the company, counting the money for each employee, and requiring the employee to come to the office to collect the salary. On their side, employees could now take their check to the bank and withdraw the whole sum, but banks persuaded customers to let the money stay in an account and to use their own personal checks for large expenses.

The new system of using checks was far more efficient and more secure than the old system based on cash. Employees also had the advantage of not having to handle large cash amounts. Psychologically it became easier to save a part of each month's salary, just by letting it stay in the bank. Checks were more secure than cash, but part of the expenses of such a system had to be paid in transaction fees.

The banks were the winners. Checks enabled them to be a part of the weekly or monthly salary transactions. Most employees of firms that paid with checks became their customers. After some years it became the norm to pay salaries using checks, even for many smaller firms. Then all participants in the workplace became bank customers. Similarly, since checks were used to pay in shops, restaurants, hotels, and elsewhere, most merchants also became bank customers. For them it was also convenient to endorse the checks and insert the money into their own account, rather than withdraw it as cash.

The problem with checks is that they can be misused. The techniques used to avoid falsified checks were the same as those used to protect paper money, such as special paper, making it difficult to alter the data after the check has been written, and watermarks. A clear weakness with checks is that they are authorized with only a signature, and that the receiver will not know if funds are available before the check is cleared.

Because checks are in paper form, they must be transported physically to the bank. Handling checks became easier when the bottom part of the check was imprinted with machine-readable characters. This opened the way for using machines to sort checks. Data other than the preprinted information, such as date and amount, had to be typed in. Later, programs were used that could recognize the hand-written numbers. Still, the physical and manual parts of check handling make it a costly form of payment. Bank of America estimates that handling a business check can cost between $4 and $20 if one includes all costs, such as the price of the check, the time used to write the check, mailing, and so on.[1] It is estimated that checks cost 10 times more than electronic payments.[2]

In clearing checks, banks are now moving images of checks electronically rather than transporting the actual paper. It has been estimated that this method saves banks more than a billion dollars every year.[3] In addition, consumers and businesses benefit from faster processing.

Credit Cards

Credit cards, in their different forms, have a long history. They were originally issued by banks and other companies to frequent customers and often used to pay at gas stations. The idea was that static information, such as the name of the customer and account data, was embossed on the card in such a way that the card could be used to make an imprint. As the name implies, the basic idea is shop now, pay later.

American Airlines introduced an Air Travel Card as early as 1934. These cards had numbers that identified the customer and the account. It took just a few years before most major airlines accepted the card. Soon, half of the ticket revenues came through these cards. A breakthrough came when the same card could be used to pay many different merchants. The first general purpose cards were Diners, followed soon by the BankAmerica card.

Bank of America faced the same problem that we have today when introducing a new technology—that is, the advantage of most so-called network systems increase as more users are on. We see this today with social systems—Twitter, Facebook, LinkedIn, WhatsApp are good examples.[4] The challenge is then to get users, initially forgetting about revenue. Bank of America solved this by issuing credit cards to all its customers in Fresno, California—60,000 in all. Having done this, the bank could persuade merchants to accept the card.

While the impression from the card took care of the static information, the amount was entered manually on the slip and was then signed by the customer. In order to verify the card and the transaction, the merchant would have to call the bank or the credit card company. Here one had to perform a manual control, sometimes certifying the credit balance of the customer. Once the transaction had been found to be in order, the merchant would in some cases write a verification code on the slip. Then the slip would have to be sent to the credit card company or, in many cases, to a company that specialized in handling credit card transactions.

The idea behind the general purpose credit cards, as well as with the checks, was to get the banks and the credit card companies into the payment loop. While a cash transaction could go directly between the buyer and seller, without involving any bank, a credit card or a check placed the bank in a central position. A clear incentive for the banks was that this offered an opportunity to offer credit with high interest rates. While some customers managed to pay the outstanding bill every month, there were many who could not withstand the idea of shop now pay later.

Advances in computer technology promised a simpler future. Starting in 1973, Visa developed a computerized verification system that significantly reduced the time it took to verify a transaction. The merchant still had to call Visa, but the answer would be given in less than a minute. The paper slip was still in use, but as we shall see in Chapter 8, credit cards could easily be transformed to a digital economy.

Today, many credit cards offer a 30 to 45-day interest-free period, after which interest has to be paid on the total debt. Most cards require that part of the outstanding debt is paid every month. Credit card interest rates are often much higher than the interest most customers have to pay on regular bank loans. These expensive loans offer a good profit opportunity for credit card companies.

Credit card issuers may offer additional benefits, such as discounts or including travel insurance. Some cards offer a cash-back system, where customers receive a small percentage of their purchases back, either as cash or as bonus points. The retailer accepting the card payment will have to pay a fee, often between 1.5 and 3 percent of the purchase amount, or even higher for some cards. While credit cards have many advantages for customers, at least for customers that do not have a large outstanding debt on the card, the retailer has been burdened with the costs. As credit cards become the norm, it is very difficult for a retailer not to accept a card. This is especially the case for the hospitality business, such as restaurants and hotels.

Technology has made it easier for companies to offer credit cards. The universal cards are still dominated by the incumbents, such as Visa, MasterCard, Diners, and American Express. However, large stores offer their own store cards that can only be used within that particular chain or store. The advantage for the store is that they get the full amount of what the customer pays. A store card is also a convenient infra-structure for offering discounts and building customer loyalty.

UPC, EAN, ISBN, and Bar Codes

Before the advent of self-service groceries, milk was sold from churns; grain, sugar, and potato from sacks; and tea from boxes. Just a few items were pre-packed. Home deliveries, which are today presented as a new and disruptive service, were also avail-able many years ago. In several countries one could have milk delivered to one's door, as well as bread and some other products. There were trucks driving around the neighborhood offering fruit, vegetables, and even fish and meat.

With the advent of the self-service store, the need for packaged products became greater, as this allowed for greater efficiency. Pre-packed products also had the ad-vantage that one could use price tags, which allowed for greater efficiency at the cash register. However, the drawback of just typing in the price here was that one had no information about what was being sold; there was no data for providing a receipt to the customer (except for the amounts), no data for maintaining inventory, and no data for statistics.

The first means of identifying a product came as early as 1948 with a system that could automatically read product information during checkout. The idea of stretching the dots and dashes of Morse code into bars with different thickness came immediately, but it took many years before equipment could be developed that could read these codes in an efficient way.

Parallel with the computer readable code was the development of a universal product code (UPC). It was clear that the success of these two developments would come when they could merge, when producers all over the world could put a unique UPC on their products as a computer-readable bar code that could be scanned auto-matically. The next step came when the data from the bar code could be used to do online lookups in computer systems, to retrieve the price of the product, to register the sale for statistics, and to do a count-down on stocks.

This is a very good example of the symbiosis between technology—represented here as bar code readers and databases with prices and product information—and standards, here the UPC system. Later on we shall see that additional technology, such as data networks, are needed to realize efficient digital payment systems.

Today, most products have an International Article Number, otherwise known as a European Article Number (EAN). Other national codes, such as the American UPC, have been embedded into the EAN. The main part of an EAN offers informa-tion on the manufacturer and product. The ISBN, the International Standard Book Number, is now compatible with the EAN. An ISBN identifies the country, publisher, and a unique number that the publisher or some other agency assigns to the book. This can be, for example, a consecutive numbering of books from 1 upward for this publisher.

Lasers are now used to scan bar codes. The devices are inexpensive and robust and the process of scanning is so simple that it is often left to the customer—for example, in self-checkout machines, a growing trend in department stores. These can scan bar codes from all directions, which leaves the customer only having to put items on a conveyor belt.

1234

Figure 6.1. The same number in Times New Roman and bar code font.

Printing bar codes is a simple task. The EAN-13 code can be found as a font that can be installed in a standard word processor, such as Microsoft Word. Numbers can then be printed in the bar code font, as seen in Figure 6.1.

Figure 6.2. QR code (example)

In some settings, two-dimensional QR codes (as shown in Figure 6.2) are used. These can code more information than a bar code and need somewhat more sophisticated scanners. Some smartphone apps can read QR codes by taking a picture of the code. Most often the QR code offers a link to a website, for example to present information or to run a process.

Radio-frequency identification (RFID) is another possibility. RFID tags may be passive or active. Passive tags are inexpensive but need to be close to the reader. Active tags have a battery in the circuit and still cost just a few cents. The advantage is that the reader no longer has to be in direct or close contact, with the tag. With RFID, one could theoretically carry a bag of groceries out the door while a scanner reads your credit card and the price of all items. In practice, it is not as simple. Even with battery-powered tags it is complicated to read items stored in a bag or in a shopping cart. However, RFID is used in cards for contactless payments (along with other technologies), in ID cards, and many other applications.

The Equipment

Automation is achieved when

1. The necessary standards are in place
2. The data is formalized
3. The equipment to automate is available

The previous section on product codes and bar codes provides a good example of automation. In order to have an efficient check-out function, every product needs a unique code. Manually identifying products by type or name is time-consuming, spelling may pose a problem and the same product may exist under several different types and names. All of these problems are avoided by having a unique product number, leaving it to the producer to decide if a new version of a product should get a new number or not.

In order to get all producers all over the world to adhere to the code scheme, it must have some clear advantages. These come when the code can be presented in computer-readable form, in our example as a bar code, and where good scanners are available. Then the stores install scanners, and then most producers put bar codes on their products and there will soon be a situation where stores will only sell products that have a code. A more formalized world is created.

As we have seen, technology is an important factor. Bar code scanners need to be inexpensive and robust. To achieve efficiency, the code must be scanned in a split second, keeping errors to a minimum. Many of the early bar code systems were abandoned because the readers, which often used incandescent light bulbs, were unreliable. Reliable scanning systems only existed when these were replaced with lasers.

All of this would have little effect on its own if the scanners could not be connected to other systems. A manual cash register would have little use for a scanner. In the beginning, computers were too expensive and too large to incorporate in a point-of-sale terminal.

In the early 1970s, Intel and other manufacturers developed a new way of making computer components. Instead of using separate electrical components, such as transistors and capacitors, and connecting these with wires, one could print the components on silicon chips. Large-scale integration (LSI), and later very-large-scale integration (VLSI), became the norm. In practice this allowed for the production of inexpensive, small, robust, and energy-efficient computer processors, the backbone of PCs, laptops, smartphones, and any computer device. Disk technology had a similar evolution. In a magnetic disk, information—in practice, binary numbers consisting of zeros and ones—can be stored on the disk by magnetizing the surface of the disk. From large units with minimal capacity in the 1960s, disk development mirrored the development of the computer; each new version had more capacity and was smaller than the last.

Now the cash register could be a computer. The scanned product code could then be transferred to the computer and used for a lookup in a database, where the name of the product and the price can be found. This could then be printed on a receipt given to the customer. The cash register displays the total. If the customer pays digitally, for example by offering a credit card to the point-of-sale terminal, all of the information needed to perform the payment is then available digitally: the credit card number, other data from the card, then the amount and date from the cash register. The cash register would then transfer all the data in a standardized format to a central clearing system.

In addition, the store will now have valuable data about what the customer bought, which would be even more valuable if the customer was identified to the store; for example, by using the store's loyalty or credit card (see Chapter 8).

All point-of-sale terminals are now online and the payment transaction, including verification of the payment card, can also be done online. This is not a problem in a society that has a good digital infrastructure. Without an online solution, however, as was the case in the early days of networking, the transaction data could be saved in the terminal during the day and then sent to the central clearing system after shopping hours, most often over a dial-up network. This "batch" system did not allow for online verification of credit cards.

The online variant offers other advantages. For example, it may be used to offer loyalty discounts directly to customers. The system will do a lookup based on the customer ID and the items that are bought and automatically compute a discount that is withdrawn from the total amount before paying. With a good broadband connection, this may be performed immediately.

Conclusion

The transition from cash to representing money in a computer is clearly not much more radical than moving from using precious metals and coins to banknotes. What we see is that the concept of money can have many representations, similar to books, music, maps, and most other symbolic systems. Digital representations today have many advantages for storing, analyzing, and transporting data.

For money, there are other advantages as well. The data can be collected in electronic form where they originate, such as in a point-of-sale terminal. These data can then be transmitted immediately to credit card companies and banks in their original form without any need for human intervention. This makes it possible to have updated balances on all accounts. It is also possible to retain all data from the transaction, not just the sum but also a list of items that were bought. In the future (it is already here in some versions) there will be no need to retain the receipt from the shop as the same data will be available online.

Notes

[1] https://www.wsj.com/articles/SB10001424052702304732804579425233344430424
[2] http://ww2.cfo.com/applications/2015/10/electronic-payments-10-times-cheaper-checks/
[3] http://www.businessinsider.com/the-death-of-the-paper-check-2013?r=US&IR =T& IR =T

[4] Interestingly, we may also have an opposite network effect—that is, when a person has so many friends on Facebook (parents, old boyfriends, teachers, and more) and get so many posts that the only solution seems to close the account.

Chapter 7
Fundamentals for a Digital Economy

Digital and cash-based economies both require trust. All participants want to be sure that the money in their wallets or accounts will retain its value. Many countries have managed to develop a national currency that is trusted in the population and is fairly stable with regard to other currencies. A central bank and a well-developed economy are essential for achieving this goal.

Trust in a digital economy can also be based on this trust in the currency. It is the same dollars, pounds, yen, or euro that are handled here. Participants also need trust in the technology, that the transactions are performed as they should be, and that the system is available when needed. However, since this is the same technology that we use for many other tasks, most citizens will already regard this as trustworthy, at least in the way that they are willing to use the technology for many different types of tasks.

I shall explore these issues in this chapter, which also includes a discussion of costs. As we shall see, the low direct cost of each transaction is an important factor in developing a cash-free society.

Trusting the Currency

A person in a country that has weak economic expectations may not trust the national currency. The answer may be to store funds in another currency, even making payments in it. US dollars and euros are often used in such cases. The person is then protected from inflation and fluctuations in the national currency with regard to cash holdings. Bank accounts will normally be in the national currency. However, some countries allow customers to have accounts based in a foreign currency; for example, a US dollar account. Value may also be stored by other means, such as gold.

Most well-developed countries have an active and independent central bank that has resources to support the national currency. The central bank will set interest rates. As we have seen, these are important in order to balance between excessive inflation or a depression. In times of crisis, the central banks have many other weapons that can be used, such as buying or selling bonds or setting reserve requirements to banks. In smaller countries that have their own currency, the central bank can try to control exchange rates by selling or buying dollars or euros. As we have seen, in many cases a weakened national currency can also be an advantage, especially to export industries. If the product you sell costs an amount of 1000 in the national currency, it will equate to $200 (based on an exchange rate of five to the dollar), but only $143 with an exchange rate of seven to one. More expensive international currencies will also raise the price of imported goods and holidays abroad. This will also boost national producers.

A negative effect of many states using the same currency, such as in the United States and in the Eurozone, is that one cannot use currency differences to boost the economy of a single state. For example, Mississippi, Romania, and other states that have a low GDP per capita cannot have their own exchange rate or their own interest rates.

However, there is no such thing as a free lunch. A weaker local currency will in practice imply lower wages; however, while it is psychologically impossible to lower wages directly, a price increase for foreign products will be taken in stride. There is also the danger that investors may have an interest in attacking a currency. They may sell the currency "short," agreeing to sell it at a lower exchange rate than the current. These attacks will be especially harmful when a national currency is pegged to other currencies, often dollars or euros.

Probably the best-known currency attack was that on the Thai baht in 1997. Thailand had pegged its baht to the US dollar, aiming to stabilize the national currency and thus stimulate the export industries. However, at that time, Thailand's government and also private citizens had huge debts in dollars. Around 1995, interest rates in the United States increased, strengthening the value of the dollar. At the same time, the economy of Thailand came under pressure. Investors sold baht to avoid losses. As a counter-attack to this flow of baht to dollars and other foreign currencies, the central bank can sell its reserves of these currencies to develop a flow in the other direction. However, the central bank of Thailand did not have these reserves.

Another option would have been to raise domestic interest rates to make it more profitable to keep funds in baht. This was done, but high interest rates weaken an economy and may cause further problems. In the end, the Thai government was forced to remove the peg and let the baht float. This started the Asian financial crisis.

Currency problems can also hit more developed countries. The British pound, the world's oldest currency, was attacked in 1992. The famous investor George Soros shorted the pound—that is, he speculated that the value of the pound would drop compared to other currencies. At that time the pound was not allowed to fluctuate by more than 6 percent with regard to currencies of other EU countries. The Bank of England tried to strengthen the pound by buying the currency. They also wanted to increase interest rates, but did not receive authorization to do this. In the end, the attack forced the British government to withdraw the pound from the European Exchange Rate Mechanism on 16 September 1992 ("Black Wednesday"). Soros pocketed a billion dollars from his efforts.[1] From then on the pound has fluctuated freely with regard to other currencies.

Switzerland had the opposite problem. Here, the danger was not a weaker national currency, but a stronger one. During the turmoil of financial markets around 2011, many investors viewed the Swiss franc as a safe haven and began to purchase Swiss francs in large amounts. This pushed up its value with regard to other currencies. A strong national currency will hurt the export industries. In order to reduce the value of the franc, the Swiss National Bank created more francs and used these to buy euros to create a flow in the opposite direction. This was successful and caused the value of the franc to fall relative to the euro, down to an exchange rate of 1.2 (one has to pay 1.2 francs for an euro). However, by 2015 many Swiss expected that the ongoing creation of francs and the large foreign-exchange reserves would cause hyperinflation. For political reasons, the central bank allowed the franc to float freely with regard to other currencies.[2] At the time of writing this book the exchange rate is 1.1, which means the franc is somewhat stronger than what the Swiss National Bank wanted it to be.

Still, while citizens in Thailand experienced a situation where their savings in baht lost much of their value, citizens in the much more solid economy of Switzerland could trust their franc. The 1.1 exchange rate will make Swiss products more expensive abroad, but most Swiss export industries can live with that.

Trusting Banks

Banks may fail, but in order to establish an efficient economy, trust in banks or the other financial institutions that handle payments and funds is a requirement. If the banks get into problems and the authorities are not willing to step in, the money may be lost for the customers, which will hurt the economy. Some people who fear this may end up hiding their currencies at home or buying gold.

In many developed countries, the government will stand behind banks. There is sometimes also a guarantee. The EU will guarantee the first 100,000 euros in an account—that is, if the bank gets into problems the state will step in and there will be no loss for customers up to that amount. In other countries the guaranteed amount may be even higher. Norway, for example, guarantees amounts up to 2 million Norwegian kroner—more than 200,000 euros.[3]

Even when there is no guarantee, the government will usually step in to save the accounts of ordinary citizens. In most cases, a loss of trust in banks will be much more expensive than the cost of saving the customers' money. A loss of trust will have serious drawbacks for the economy. In the panic that will follow such a situation, citizens may be more careful about using money and will store money at home, thereby reducing its effectiveness. That is, independent of explicit guarantees, governments will do everything possible to avoid ordinary citizens losing their savings.

A digital economy needs a similar level of trust in banks and the other financial companies. As discussed above, there is also a need to trust the technology. While this may be the case in normal situations, computer errors, network problems, and the loss of power may have serious consequences as it may not be possible to perform electronic payments or to access Internet banks. If such errors persist, the population may lose its trust in the technology. We shall return to this discussion in Chapter 10.

Transaction Fees

While customers and retailers would like zero transaction fees, any type of transaction incurs expenses and these expenses must be covered. The customer can pay an explicit amount for each transaction, or the merchant can pay, or the expenses can be shared between these parties. Alternatively, as is the rule with cash, the transaction expenses can be hidden among all other expenses.

Ideally, if there is a transaction fee it should be borne by the customer. Then the customer can take this into account when choosing how to pay. The problem is that we expect to have zero fees when paying in cash. There are traditional reasons for this.

Even when a cash transaction is the most expensive option for the store, it will be very difficult to ask the customer to pay a fee. As discussed in Chapter 4, cash becomes expensive when it stops being agile. Then the cost of counting, storing and transporting the cash becomes prohibitive. This is especially the case when only a few customers use cash. The cost per transaction may then run into much higher percentages than the fees for a credit card transaction.

Debit cards, where the amount is retrieved directly from the customer's bank account, may have transaction fees of only a few cents per transaction. Since there is no credit involved, no risks, no cash back, no insurance, or any other perks connected to these payments, they can be performed by a simple, low-cost, computer transaction. However, the bank may charge the customer an annual fee for the card.

Credit cards are more expensive to use. A store will usually have to pay a fee of between 1.5 and 3 percent of the total amount, higher for some cards. Since competition is tough between the various providers, the credit cards will often be offered free of charge to the customer. In many countries, credit cards have zero transaction fees for the customer; in others, restaurants or stores may warn the customer that the transaction fee will be added to the check. Usually this will be in the form of a small percentage.

There are also examples where the fee added to a credit card transaction is much higher than what the retailer has to pay, discouraging the use of digital payments. For example, a UK taxi company warns that it will add 10 percent to the bill if a credit card is used. One can speculate that the idea here is to keep income in cash, which has several advantages if the company wants to avoid paying VAT or other taxes.

Some banks charge their retailers a fixed minimum fee in addition to the percentage when a credit card is used. Merchants then feel obliged to introduce a minimum amount for using a card. This is not efficient. For the customers it will be convenient if the same payment option can be used everywhere, independent of the amount. In this case we see that the fee structure of the banks is maintaining the use of cash. In the long run it will be difficult to discourage digital payments for small amounts. When customers can pay small amounts conveniently by using their mobile phone or a tap-to-pay card, they will not accept high fees or resort to cash in these cases. Also, when the technology is in place, the cost of yet another transaction is very low.

Some countries do not demand any fees or minimum amounts, which makes it convenient to use digital payment for just about all transactions. As we shall see in the case study in Chapter 14, this may be a practical requirement for establishing a cash-free payment system.

Since it is difficult to put explicit fees on the usage of cash, digital payments should also enjoy zero fees on payments. This can be achieved either by asking the merchant to pay the 1.5-3 percent of the total for credit card payments, or by having a zero fee on debit cards. In the latter case, costs will be negligible.

Digital for All

The fundamentals for establishing a digital economy are that consumers trust the currency and the financial institutions. We discussed these issues above. In addition, there are several practical issues. Every consumer who wants to participate in the digital economy in the developed world needs a bank account or credit card account. In some countries banks will only offer accounts to customers who meet a set of requirements, such as a regular income and a place of residence. In others, regulations may stipulate that banks cannot refuse to set up an account. Usually the default account will be one that can be accessed with a debit banking card. Since it is not possible to withdraw money that is not in the account, the risk for the bank is minimal.

While an account with a bank or credit card company is the entrance requirement to the digital economy in a developed country, several underdeveloped countries offer banking services through mobile phones. Customers are then identified by their SIM cards.

If the complete population is not offered the possibility of participating in the digital economy, it will be necessary to maintain a cash economy in parallel with the digital. This can be expensive, especially once most transactions are performed digitally. But since the technology offers the possibility of checking the balance before a withdrawal, it becomes risk-free to offer a bank account to every citizen. We see that the digital economy solves the problems that it creates.

An alternative is to offer cash cards. In some countries, social security support may be offered in this way. A cash card can be used as cash, but the notes are here replaced by a plastic card with an amount on the card. This can be used in all terminals, similar to a banking or credit card. These cash cards may also be an option for small children. Parents can then fill up the card from their own accounts. However, in some countries, even children are offered standard debit cards. This seems to function very well. Experience shows that 10-year-olds are even better at protecting their cards than teenagers.

Most tourists have international credit cards, enabling them to participate in the digital economy in the visiting countries directly. Debit cards are often national and will only work in the country where they are issued. Tourists without a credit card rely on cash, either in the currency of the country that they visit or in a well-known currency, such as euros or US dollars. In a digital society these tourist would have to change the representation of their money from cash to a cash card. In the long run, we should expect most tourists to travel with an international credit card, which is also needed to rent a car or to check into many hotels.

A digital economy will require that the technology, from terminals to computer networks, is in place. It is also necessary to have a population that can master the technology, log in to online banking, perform basic monetary transactions, and more. This will be made easier as the technology moves forward. For example, smartphones offer good opportunities for simplifying payments. This will require the ability to download, install, and use apps. However, this is similar to installing apps for other purposes, which most people manage to do.

Since banks and credit card or cash card issuers are the backbone of a digital economy, these organizations need to go digital and be trusted by the population. If banks can fail and if there is no central government to protect customers, many people will feel more secure using cash, especially dollars or euros.

As we shall discuss later, several international companies such as Facebook, Apple, Amazon, Google, or Alibaba are in the process of establishing digital payment systems that can be used by everyone. However, for a citizen who receives his or her income in an unappreciated national currency, it may not be a viable option to use digital payment systems based on other currencies.

Conclusion

A digital economy must be based on a well-functioning currency, and banks and other financial institutions must play an active part. It will also be necessary to have the technological infrastructure in place, everything from terminals to data networks, central servers, and clearing systems. We shall discuss these aspects in detail in the next chapter.

Another requirement is a population that is comfortable with using digital systems. This will be the case in most developed countries, especially as the new payment services will be well integrated with other digital tasks, such as booking tickets or shopping online. Further, it is a requirement that all customers are in a situation where they may have a bank account. If not, expensive cash services would have to be retained for a small fraction of customers.

Zero fees will be important in leading customers from cash to digital forms of payment. If not, the tradition of zero fees for cash transactions, even where cash is more expensive to handle than the digital payments, may hinder the development of a cash-free society.

Notes

[1] Mallaby, Sebastian (2011) *More Money Than God: Hedge Funds and the Making of a New Elite*, Penguin Press.

[2] https://www.economist.com/blogs/economist-explains/2015/01/economist-explains-13

[3] With an exchange rate of 9.6 kroner to the euro (one has to pay 9.6 kroner for one euro). This is the rate that we shall use in this book. The EU has criticized Norway for having a higher government guarantee, claiming that this makes Norwegian banks more competitive.

Chapter 8
Infrastructure for Digital Payments

To use digital payments it is necessary to have a digital infrastructure, point-of-sale terminals, computer networks, and clearing systems for handling transactions between banks, credit card systems, and so on. These are all discussed below.

Terminals

In a digital economy, all points of sale must be able to take a digital payment. Today we achieve this by having a terminal that may be connected to the cash register. Most terminals can accept cards with a magnetic stripe or cards with a chip. Newer terminals will also be able to handle cards with an embedded RFID[1] chip for tap-to-pay functionality. With NFC[2] and Bluetooth[3] radio communication, the terminal can also connect to smartphones, either for mobile payment or tap-to-pay functionality.

Providers of terminals and payment systems know that retailers will not install systems that can be used by only by a few customers; nor will they want to have many different terminals lined up at the cash register. This forces the providers to cooperate, either by agreeing on a standard or allowing for one terminal to handle different systems. On the other hand, customers want to choose how they pay and will expect that the merchant can handle different credit cards or different electronic payment systems. This requires cooperation between banks, credit card companies, and retailers. However, experience from many countries shows that this is achievable; the customers can get flexibility and the retailers will get a one-size-fits-all system.

Terminals need to have a network connection to be able to verify cards and transaction amounts, and also to transmit the data from the transactions. Some retailers also use this connection to compute discounts—for example, based on the items the customer has bought. While such two-way communication was impractical with low-speed networks, it is now achievable with broadband connections. It is, of course, important that payment does not become a bottleneck.

We traditionally think of a payment terminal as something that is installed at the retailer, but this can just as well be a device that is owned by the customer, such as a smartphone. This opens up many new possibilities, including very simple identification and payment systems. I explore these possibilities below.

Smartphones

The smartphone has several advantages for making payments:

1. The customer buys, maintains, and handles the equipment, both hardware and software.
2. The customer is normally an experienced user of the device.
3. The software can conveniently be installed as apps.
4. The smartphone is always online and can be connected to websites, email, and more.

5.   The ID mechanisms of the phone, fingerprint reader, iris scanner, face recognition or passwords, can be used, sometimes avoiding another set of identification.
6.   Phones have a display that can be used to select items, to offer additional data, and to provide forms for input.
7.   The location mechanism can often be used to simplify data entry.
8.   A receipt, ticket, or similar can be stored on the phone.
9.   The receipt can be presented in many different formats, including those that can be read digitally.

Companies are eager to develop new apps that utilize these advantages. For example, Ruter, the company handling local area traffic in Oslo, developed a ticket app that turned out to be highly successful and has been downloaded more than one million times. According to Ruter, more than half of Oslo commuters used the app in 2017, giving Ruter one of the highest mobile phone usage percentages in the world.

The app utilized all the advantages mentioned above. The customer maintains the equipment (1), has user experience (2), downloads and installs the software as an app (3). The app can be used everywhere and receipts can now be stored on the phone, on a website, or sent as email (4). After setting up an account, registering the name, credit card, and so on, access is given directly by using the ID mechanism of the phone (5). Tickets can now be bought just by entering the zone or destination in a customized user interface (6). By using the fact that phones can be located, the input part can be simplified by automatically suggesting the from-place (7). After the ticket is bought, the actual ticket is stored on the phone with a countdown mechanism to show validity (8). The ticket is presented as a QR code that can be read digitally when entering a train or a bus, or by a screen that one can show to the bus driver (9).

With smartphone applications such as this one, the payment is an integrated part of the system. This will be an important enforcer of a cash-free society. Digital payments are no longer just a replacement of the previous cash transactions, but offer something very different. Now the actual payment may be just a part of the whole transaction. In the example above, the transaction has many parts: finding a route, retrieving the time of departure, choosing a ticket, and paying. As we have seen, the actual payment is just a small part, often performed just by a button click.

Now smartphones are also being introduced as a way of paying at a terminal—for example, at a retailer. Then the NFC capabilities of the terminal may be used, using the phone as a tap-to-pay card. Other possibilities are that the store offers a QR code or a phone number for identification. In the first case, the customer will use an app to read the code (by taking a picture); in the other one, he or she enters the phone number to identify the receiver. Both systems require no installation on the supplier's side. This makes these systems convenient for small retailers, such as those selling local goods at a market booth or, for example, paying for parking or tolls in rural areas.

However, without good Internet coverage, the smartphone will be impotent. While some countries have coverage in all population centers, there may be others with only sporadic coverage. We should expect that this situation will improve. The functionality of the smartphone will in itself generate pressure for better coverage. We will come back to these forms of payments in Chapter 9.

Computer Networks

In order to get online updates of accounts, which are important both to give a good overview of transactions and to avoid overdraft, point-of-sale terminals of any type need to be online. This will be most efficient when the terminals are on a broadband computer network. This is becoming the standard today in most modern countries. Telephone lines and coaxial cables are being replaced by fiber-optic cables that offer a significant increase in bandwidth. The advantage for the customer and the store is that digital payments can be performed very quickly, in most cases much faster than cash transactions. As we have seen, high bandwidth also opens the way for new applications, such as those that can compute discounts immediately.

Mobile networks are also moving toward greater bandwidth. Thus, a payment using a smartphone may also be expedient, at least where there is good network coverage.

Clearing Systems

Transactions that are received from point-of-sale terminals and smartphones must be sent to the correct bank or credit card broker. This is performed by a clearing system. Countries that have a good national system will have a clear advantage in this regard. Then the transaction can be forwarded to any bank or credit card issuer, independent of where it originated. That is, there is no delay in the transaction, even if the customer and store use different banks.

In some countries, such as the United States, an account number is unique only in the bank that created the account. This makes it somewhat more complicated to perform bank-to-bank transactions. The Automated Clearing House in the United States is considered an anachronism by global standards.[4] In many other countries an account number is unique on a national level, making it unnecessary to add additional information for bank-to-bank transfers. Several countries, such as the Scandinavian, also have a common clearing house for all banks.

Digital Credit Cards

As we saw in Chapter 5, early credit cards were used to make an impression on a paper slip. Although they had much in common with checks, an advantage was that the customer did not have to bring a check form, as the retailer had the necessary paper slips. Of course, the cards also offered credit.

The data on the card consisted of the customer's name and account number and also the expiration date. The amount had to be filled in by hand, the slip had to be signed, and then the transaction had to be verified by calling the credit card issuer. Then the slips had to be transported to the bank and the transactions registered in the various accounts.

All of the data is now formalized, as are the processes. With the high number of transactions, this is a system that is the ideal application for modern computer technology. As we have seen, inexpensive terminals and reliable computer networks enabled this development. The data on the credit card, now with a magnetic stripe or a chip, can be read automatically. The amount can be typed on the terminal or, even better, received directly from the cash register. Verification can be performed online in a transaction between the terminal and a central server, and all the further processes can be performed automatically based on the digital data.

Signatures, which are not easy to incorporate in a digital system, have been replaced by a PIN code. The advantage is that the PIN can be offered using any type of keyboard and the verification is a very simple process. With signatures, special tablets would have had to register the signature and a complex process would have been required to compare this to the signature registered at the bank. In practice, such a process would not work. The alternatives would have been either accepting all signatures that looked somewhat like the original, with the possibility of accepting an illegal signature, or risking that legal signatures were not accepted. If we return to our ideas of formalization, we see that while the PIN code is formalized for an operation such as verification, the signature is only partly so.

In a digital credit card system with online verification, humans can be taken out of the loop. This results in an efficient system that can handle a very high number of transactions, in practice many thousands per second. Transactions can be handled immediately so that, in principle, all accounts can be updated at any time. Digital systems also provide a set of statistical data that will provide valuable information for all stakeholders.

Digital Debit Cards

A debit card withdraws the transaction amount from the bank account directly, which means the customer must have the necessary funds available in the account. The advantage for the store is that it gets its money immediately and the transaction fees are much lower than for credit cards.

In some countries the fee on a debit card transaction may be as low as a few cents (see Chapter 14). With broadband terminals and automatic processing, users can take advantage of the speed of modern computer technology. Because there is no credit and usually no other perks, costs are kept to a minimum. In a digital society, the debit card transaction may be the default, with no fees for the customer, while users will have to pay additional fees for the more expensive variants, such as credit cards or cash.

Cash Cards

A cash card is a plastic card that can be used in similar ways to debit or credit cards. It can be bought in fixed denominations. The main difference is that it is usually anonymous for the consumer as the card does not represent the owner. There is no connection to any personal banking account, which means the cards can be used by anyone, similar to cash. If a card is lost, the value on the card is also lost, just like cash. However, cash card transactions are not anonymous on the part of the retailer. The amount from the transaction will go into the retailer's account, similar to payments with debit or credit cards. In this way there will be less possibility of tax evasion on the seller side than if cash had been used.

One advantage of cash cards or any type of stored value cards is that there is no need for any external connection. The value is stored on the card, in many ways similar to cash. These cards are normally used for low-value transactions, such as public transit or phone calls. Some of these cards can hold various currencies. The advantage is that the user gets the exchange rate at the time the card is loaded.

Criminals have seen the advantage of these anonymous cards: they can be bought with cash and offer the same anonymity as cash. Some financial institutions allow customers to transfer large amounts from an account to an anonymous cash or credit card. While it is not impossible to trace the use of these cards, the anonymity makes it much more difficult. There are already examples of criminals using this option to launder money using banks in other countries to perform the transfer.[5]

The anonymity has other drawbacks. When crypto-criminals have hidden all data on a victim's PC, using a virus that runs a cryptographic algorithm, they may demand ransom money in the form of anonymous money such as cash cards or bitcoin (we shall discuss bitcoin in Chapter 11).

Anonymity for small transactions can be achieved by a "wallet" inside a credit or a debit card, instead of using a separate card. This wallet could be loaded with money from the account and then used as a cash card. The balance would be given at any time, but there would not be any registration of each item used. Cash cards or the "wallet" equivalent can be a way of giving money to children. In many countries, children aged eight or older can receive a debit card on an account, but younger children may get a cash card that can be loaded with limited amounts from the accounts of their parents.

Security

A digital economy must be protected. In some ways this is easier than with cash-based systems. Cards can be protected by chips that make them difficult to copy. Usage can be protected by PIN codes and fingerprint scanners. Online systems where transactions are recorded as they are performed will also make the system more secure (see Chapter 15), and with debit cards and online systems it is not possible to use money that one does not have.

Many modern banking systems use smartphones for additional authentication and increased security. With both phone and PCs we get what is called a two-pass system. For example, a customer may be in the process of paying on a website. She has typed in the credit card number, the month of expiration, and her three-digit credit card verification code (CVC). The next step will be handled by the credit card issuer. A message with a keyword (the same one offered on the website) will be sent to her phone. After she has confirmed that the two keywords are identical, she will be asked to type in the PIN. That is, in order to perform the transaction, she needs her credit card details and access to both the smartphone and the PIN.

By hacking consumers' PCs and smartphones, criminals have been able to break security systems as complex as these. In fact, it is worryingly easy to fake emails or hack a PC. We still use the SMTP email protocol—the simple mail transfer protocol in which it is quite easy to fake the sender's address. While the text of a link in an email may appear plausible, the actual link may send the user to a very different address. Also, many users are tricked into opening dangerous attachments.[6]

When even large companies, such as Yahoo and the credit-monitoring firm Equifax,[7] are vulnerable to hacking and virus attacks, one may wonder whether it is really possible to make a secure computer system. The complexity, many layers of code, and large number of code lines make this difficult. The main problem is that we want these systems to be accessible over the Internet. If such access were not required, we could put the computer in a locked room with no external lines and no wireless. However, we would then not be able to perform many of our daily operations—no email, no Facebook, no Google.

As was the case with cars, security for computers was something of an afterthought. The first cars went on the roads at the end of the nineteenth century. It took seventy years before car manufacturers and authorities took the problem of safety seriously. Similar to the development of cars, the idea for computer and software manufacturers has been to implement more functions rather than allocate resources to security. However, after some spectacular attacks, where personal information from millions of users has been stolen, and some smaller-scale attacks where private individuals and small companies have had all their data made inaccessible, the problem is now being taken seriously.

This book discusses payment systems. Payment always involves money or its corollaries and money has always been under attack from criminals, from armed bank robbers to muggers and counterfeiters. Digital payment systems will also be under attack. In February 2016, hackers tried to steal 951 million US dollars from Bangladesh Bank and managed to get away with more than 60 million dollars. An important final step in all computer fraud is to convert the funds into anonymous cash at one point or another. Without the cash option, criminal activities like these would be more difficult or at least easier to investigate after the crime.

While the incentive for attacking a bank or a payment system is clear, criminals also attack personal PCs and the computer systems of organizations. A common form of attack is getting users to open an attachment with malware. When executed, this may run a cryptographic algorithm on all data that is accessible to the virus program, including data on the main disks, disks that are connected to the PC, or disks that are accessible over the local network. The criminals will then ask for payment in order to provide the victim with the key they need to decrypt the data. Payment may be demanded as anonymous cash cards, but today asking for payment in a currency such as bitcoin is common. The best remedy against this form of attack is to have updated systems and offline backup, and to offer clear warnings to users to be careful when opening attachments.

However, many of these emails look genuine. They are the bait in this "phishing" attack, where the idea is to get confidential information such as user names and passwords, or to lure the user into performing transactions that benefit the criminals. This is often achieved by directing the user to fake websites that look and feel genuine.

The email notifying you of the agenda for an upcoming meeting may have an attachment. If the criminals are smart, and use time to prepare an attack on your company, there may be little difference between a genuine email and a fake.

Of course, most of these fake emails are mass-produced and sent to millions of email addresses with the hope that some will open the attachment. A common strategy is to make these phishing emails look genuine; for example, as a note on a parcel delivery from Amazon, or asking to reactivate or confirm a PayPal account, hoping that the receiver is expecting such an email. Another strategy is to offer phishing messages that most users will detect immediately as a fake, such as "click here to receive your $1000 bonus." The idea may then be to reach those willing to follow up, in the end to pay the $100 transaction fee that is required to clear the bonus.

It is a pathetic commentary on our computer professionals that email systems fail to protect users from such simple-minded attacks. For example, if there is a difference between the displayed link and the actual link, color coding could be used as a warning. Similarly, dangerous attachments such as zip files could be displayed in red. Further, any attachment that looks suspect should be opened in a secure "sandbox"— that is, in a program that displays the contents without invoking any code. Many "ordinary" people, including elderly people, are now amateur computer users and need to be protected.

Offering a Continuous Service

Digital payment systems can go down. Terminals may stop functioning and may lose access to the network. Problems in a power grid will have dramatic effects. Without power, many other functions will be unavailable.

A modern society is dependent on a robust and stable power supply. The only way to achieve this is to build redundancy into the system—that is, several power stations and several transmission lines. If one fails, there should be another for backup. It is problematic if many power networks are at full capacity and the business model in many countries does not encourage companies to invest in redundancy or a high level of maintenance. Therefore, there is little room for handling unexpected situations. The remedy is to invest more, but this probably needs government interference as the power companies may not be willing to pay extra for redundancy.

New battery technology may offer an excellent local power backup, since this can operate independent of other systems. Again, however, the technology is expensive and companies may not be willing to prioritize this.

Similar issues are apparent with computer networks and payment systems. They can also be made more robust by redundancy. Security can be achieved by having complete backup systems. While this may be an expensive solution, backup can often be established by agreements between the competitors in a market. For example, if one operator has problems with its mobile network, customers may be allowed to use the competitor's networks, perhaps with a limit on the data and telephone usage to avoid congestion.

As with the power networks, the operators cannot be expected to be willing to accept the costs of good backup systems. One possibility is that the authorities fine companies for disruption in their service. With heavy fines, the companies may find it more economical to invest in reciprocal agreements with competitors or install backup systems.

Intermediate Systems

In the transition from one technology to another, we often find hybrid systems that simplify the transition. Fax machines are a good example. These were widely used 30 years ago and were based on a sheet of paper. Then the only requirement was that both parties had a fax machine connected to the telephone network. While the quality would be poor and the transmission fairly slow, an advantage was that the method relied on the paper standard. Anything that was on a sheet of paper could be transmitted: printed or handwritten text, diagrams, drawings, pictures, and so on. As long as a document could be printed, it did not matter what kind of word processor, if at all, was used.

While the fax machine was a godsend in the early 1980s, it has fallen out of use in most places today. By sending documents as attachments to emails, or submitting them to websites, material can be transmitted immediately in high-quality form. This option requires that the relevant standards are in place and that the receiver has the software to read the documents in the right format, such as a PDF or a Microsoft Word file. There is no longer a need for fax machines in most settings; this was an intermediate system, a link in the evolution from letters to digital transmissions.

The fax machine can be compared to another intermediate machine in the automatic teller machine (ATM). ATMs are useful when cash is the norm and where retailers are still cash-based. That is, the ATM is a connection from the digital back to cash. Some advanced ATMs are also able to receive cash. However, once every store starts accepting cards, the ATM will lose its purpose. It seems foolish to move cash from the ATM to the store when the card can just as easily be used in the store. The only need for cash will then be person-to-person transactions, but as we shall see, several new digital systems are also offered to handle these operations.

Therefore, ATMs are disappearing in countries with a digital economy. It was an intermediate technology that is being replaced by further advances in digital payment systems. Also, ATMs are expensive to run and maintain. Quite sophisticated mechanics are required in order to offer the right amount of bills. The machine needs to be loaded regularly, and all procedures need to be foolproof, since money is involved. There is also a need for protection, both to protect the customer's card from criminals who try to capture the information on the card by installing readers on top of the card slot in the machines, and the ATM itself. There have been several spectacular attacks on ATMs where the criminals have used explosives or heavy equipment to get at the cash. We shall study the decline of the ATM in Chapter 14, where we present the case of a country that has moved far in the direction of a digital economy.

There has been an interesting development in many African countries, which may lack a good banking infrastructure but have a good infrastructure for mobile networking. This is used for deposits, withdrawals, transfers, and for payment of goods, all the functions that we require of a banking system. An example is M-Pesa, a mobile phone-based money transfer system in which customers can deposit and withdraw cash from a network of agents, often the shops that sell airtime. Here the shops perform the same function as an advanced ATM, both supplying and receiving cash. Amounts are transferred to other persons,; also sellers of goods, by using a menu on the phone. In Kenya, M-Pesa is used by more than 17 million Kenyans, two-thirds of the adult population.[8]

The system has been extended to offer loans and saving accounts. Salaries can also conveniently be paid through M-Pesa. Offering a good and inexpensive payment and "bank system" to a developing country where the traditional banking systems were missing seems to have positive side effects. More money is now run through the open economy, and start-ups now have a base on which to build their business.

Something similar is happening in Myanmar, where fewer than one in ten people have a bank account, but close to 90 percent have a mobile phone; many people in that country use Wave Money, a service that processes 100,000 transactions a month.[9]

Conclusion

In this chapter I have presented the infrastructure for digital payments. In most developed countries this is already in place, augmented by the fact that smartphones also may be used for making payments. This implies that one can perform a digital payment at any place, perhaps as long as there is a network connection. A smartphone also provides backup—for example, if the terminal at a store does not accept your credit card. It seems that the phone also will be the basis for a digital economy in developing countries.

Notes

[1] Radio-frequency identification (RFID). Cards with an RFID chip can be read just by holding the card close to the reader.
[2] Near-field communication (NFC). A set of communication protocols using radio transmission between two devices, such as a point-of-sale terminal and a smartphone.
[3] Bluetooth is a wireless technology for exchanging data over short distances.
[4] Rogoff, Kennet (2016) *Curse of cash*, Princeton University Press.
[5] *Nasjonalt Tverretatlig Analyse- og Etterretningssenter* (NTAES), Nyere Betalingstjenester, 2017 (in Norwegian).
[6] Even some banks have not understood how vulnerable email is. I have received emails from banks with a "log-in here" link. This is highly vulnerable as it enables others to copy the email, change the link and the send it as spam, hoping to cheat some customers. Banks should always ask customers to use the standard means of accessing their web systems, by typing the URL, searching on Google, or by inserting a shortcut icon.
[7] Learning the lessons of Equihack, *The Economist*, September 16, 2017.

---

[8] *The Economist*, May 2, 2015.
[9] *The Economist*, October 14, 2017.

Chapter 9
Digital Payments

In order to perform a digital payment in a situation where we are physically present, such as paying in a store, we need technology to

- identify the customer, or more specifically, the customer's account
- register the amount and date, and the ID of the shop
- transfer these data to a central system, such as a bank or a credit card broker.

I shall discuss this form of payment before moving to other forms of transactions.

Paying in a Store

Consumers in a store need some form of electronic ID to identify themselves, or more specifically, the account that they will use for the transaction. Credit or debit cards with a magnetic, computer readable, strip have been the norm in this regard. The information on the card is coded on the magnetic strip, and the card can be read in a simple magnetic reader. This is an old, robust, and proven technology that was used in magnetic tapes and cassettes.

A card with a magnetic strip is cheap to produce, but has the disadvantage of being easy to copy. Many types of credit card fraud have been based on making illegal copies. This can be done manually, such as in a restaurant if the waiter has your card, or electronically by mounting illegal card readers in front of the card reader slot in an ATM.

More secure cards use an embedded chip. This is much more difficult to copy and the information on the chip may be better protected than on the magnetic strip. Cards may also have an RFID[1] chip. This can be detected simply by tapping the card on a reader. Such tap-to-pay systems often allow small payments to be performed without a PIN.

With only the card as identification we have single-factor authentication. This is very convenient to use, but authorities often set an upper limit on the amount that can be paid, for example £30 (approximately US$42) in Great Britain (in 2018). For larger amounts, two-factor authentication, such as the card and a PIN, is needed. Later on, we shall offer suggestions for systems that combine simplicity and security.

As we have seen, smartphones can also be used for payments, replacing the card. Many smartphones have a convenient fingerprint reader, which makes it possible to authorize a payment without the PIN. In the simplest system, a customer pays by holding the phone next to the terminal after starting an app. Communication is then based on NFC. As the name (near-field communication) implies, this is a form of radio communication that works within a range of some centimeters. Most smartphones and many point-of-sale terminals support NFC. Several providers, such as Apple and many banks, now offer smartphone payment systems.

Other information needed to perform a payment, such as the amount and the ID of the store, can be retrieved automatically. The only data that has to be entered manually is the PIN, and this may also be avoided with a fingerprint reader.

For customers, an advantage of digital payments is that all data are registered and can be presented in a bank statement on paper or electronically. Today most systems offer limited data, such as the date, the name of the store, and the amount. However, we should expect more in the future, such as information about the items we have bought. Some systems are already available and I discuss the merits of such systems in Chapter 15.

An advantage for the store is that the payment can be registered in a customer's profile. The store will then need to connect the transaction to a customer. There are many ways of doing this. The customer can have a separate loyalty card or have his or her profile connected to a credit card. By registering all transactions, the store will learn more about each customer and can use this information for marketing, special offers, bonus points, and so on. I also discuss these aspects in Chapter 15.

The advantage of a digital payment for the bank and the credit card company is that they are now in the payment loop. In addition, the data provided by the payments will offer important insights into the customer, the retailer, and commerce as a whole. For example, such data should make it possible to build good customer profiles, information that may be used for marketing.

To date, such data has been proprietary for the banks and credit card companies. However, new EU laws, the Payments Services Directive (called PSD2), will force banks and other companies that store data on private individuals to share the data with third parties. The idea is to let customers have ownership of their own data. Since these data are confidential, the third parties have to be confirmed and the customer must explicitly give consent. This effort shows that the EU feels that information of this kind is too important to let the banks exploit their monopoly.

Commuter Tickets

Figure 9.1. App for commuter tickets (Ruter, Oslo, Norway)

There are many ways to pay for commuter tickets: on the bus or the train, in shops, or at vending machines. Different countries and different cities have had their own systems. Until recently, vending machines were seen as the future. The Oslo-based company Ruter spent more than 600 million Norwegian kroner (US$70 million) on an automatic ticket system with card readers and ticket automats at every station. The system never became operational. Then they spent 20 million Norwegian kroner (US$2.4 million) developing the app mentioned previously that became a great success (Figure 9.1).

Figure 9.2. One-click payment for an airport express train.

Apps are now the norm in many countries. With a registered user and a registered credit card, a ticket can be bought by just by a few button clicks. Some systems will also detect your location and can then automatically insert some information, saving button clicks. For example, when I buy a ticket for the airport express train at Oslo Airport (Oslo Lufthavn), it will suggest a ticket from the airport to the city center (Oslo S), the latter information having been collected from a previous trip (Figure 9.2). It is then literally a "one-button click" operation that is initiated with the "buy" ("kjøp") button.

The apps may also tell users when the bus or train will arrive according to the schedule, or better, according to actual online infomation. For example, if the bus is delayed in traffic, the system can estimate an arrival time based on distance and speed, or based on a big-data analysis of previous data.[2]

The Ruter app offers customer a set of functions that were never possible when using cash or a travel card. The app counts down to when the train or bus will depart. It can give warnings when a monthly pass is about to expire. With an app, a user can buy tickets for more than one person or buy a ticket and forward it to another person. Since the company and its customers rely on smartphones, USB contacts are now being installed in buses, trains, and boats so that passengers can charge their phones.

What we see here, and what we have discussed previously, is that the actual payment is part of a process that helps users fulfill all of their needs. This is where digital systems have a large advantage. The payment is an integrated part of the whole process, not separated out as when using cash.

Digital Payments for Net Shopping

While cash is an alternative when we are physically present in a store, it cannot be used online; here, credit cards are the most common form of payments. Alternatives are to prepay for products by transferring the amount to the store's bank account or to use a collect-on-delivery payment. However, both of these alternatives are cumbersome and collect-on-delivery may also be expensive. Many delivery services also require a digital payment. In addition, an advantage of using a credit card is that the credit card company will guarantee the transaction; if you don't get what you ordered, they will cancel the payment. In practice, net payments cannot be performed with cash.

By verifying the transaction through the use of a smartphone and PIN code, as we discussed in Chapter 8, these transactions can be made secure, at least secure enough to make customers confident that the transaction will be performed as it should be. In addition, computer systems can check for fraud. For example, the transaction may be found to be suspect if the goods ordered are expensive, easy to sell, and not being delivered to your home address. Based on big-data analysis, a computer system can detect fraud or at least indicate where fraud can be a problem. Such a system can also learn from experience, so that data on every faulty transaction can be used to improve the detection of this type of transaction in the future. There are many "machine learning" applications that have succeeded in this area.

In developed countries, most customers shop on the net. Online shopping may comprise as much as 15 percent of all shopping in some countries, even higher for some products, and the percentage is increasing every year. Nearly everyone uses the Internet to book airplane tickets and hotels. Customers who use digital payments on the Internet may want to use the same means of payments in retail stores. In this way we see that online shopping is defining digital payments as the norm.

Other Forms of Digital Payments

The advantage of using smartphones for payments is that the customer brings his or her own device, meaning there is no need for a separate point-of-sale terminal, although good Internet coverage is a requirement.

In the simplest situation, the terminal can be replaced by a QR code, an optical code that can be read by a smartphone. For example, the code could be placed in a parking area. It will then give information on the parking company, the receivers of the funds, and the amount involved. To pay, the user will open the correct app, provide the PIN code to get access and then take a picture of the QR code. Additional information, such as the license number of the car, can be typed in.

Figure 9.3. Using a smartphone for person-to-person payments.

Smartphones are also used for payments between private persons. One way of doing this is to use a bank app to get access and then transfer the amount to the receiver's bank account. The transactions can be simplified by using the receiver's phone number as a replacement for the bank account number. An example is shown in Figure 9.3. The advantage is that phone numbers are more easily recalled. Everybody has their friends' phone numbers, but few have their account numbers. It is also more convenient and more secure to offer your phone number to other people than a bank account number. This requires that all users connect their phone number to a credit card or a bank account.

Until these systems emerged, cash had the advantage of offering real-time clearing of transactions—that is, when the customer handed over the cash to the seller and received the product in return. These new phone-based systems also perform transactions immediately. That is, there may be some time before the money is taken out or inserted in the underlying accounts, but the transaction is registered in real time and cannot be revoked. Internet banks are also starting to offer real time transactions. This will probably be the norm in a few years. There is no good reason why monetary digital transaction cannot be performed immediately. The advantage is that customers receive real-time clearing and balances that are updated after every transaction.

These systems can also be used for payments in stores and for paying invoices. In the latter case, the invoice will be sent to the customer's smartphone and can be paid by a button click. No additional information has to be entered. However, the simplest way of paying with a smartphone in a store is to use NFC and the tap-to-pay service, as described above, since this requires less input.

When customers have smartphones, stores can offer apps that handle the complete customer experience. This also allows for the implementation of more complex discount systems. For example, a Norwegian grocery chain has introduced a system where customers get an immediate discount on the ten most bought products (based on frequency and value). In the future, this process of collecting data and offering discounts may be seamlessly integrated with digital payments, but not with cash.

Conclusion

In this chapter we have explored different ways of making payments in retail stores, for getting commuter tickets, and for online shopping. We have also looked at other ways of making payments, using QR codes and apps on smartphones. For many of these operations, cash either cannot be used or would be very cumbersome. We also see that the payment is only a part of a large digital transaction where all the operations are an integrated whole.

While the smartphone is ideal for many types of operations, it will not necessarily replace debit or credit cards. In a store, it may be more convenient to use the card than a phone; when verifying an online transaction the smartphone is often convenient as it offers a high degree of security. We shall see later that phones and cards may also work together in other ways.

Notes

[1] RFID–Radio-frequency identification.

[2] Several years ago I was involved in a project that estimated the time of arrival for passenger boats on the northwest coast of Norway. By collecting all data on each trip, the system was soon able to offer detailed estimates based on a big-data analysis of previous trips. See Olsen, K.A., Indredavik, B. (2011) "A Proofreading Tool Using Brute Force Techniques." *IEEE Potentials*, 30(4).

Chapter 10
Internet Banks

Banks were early adopters of new Internet technology when it emerged in the 1990s. This was a natural evolution. Banking operations are simple, the data is formalized and compact, and much can be achieved with a simple text-based interface. In this chapter I discuss the various aspects of Internet banking.[1]

Trust

Trust is important for all banks. While the bricks-and-mortar bank may indicate trust by having a large stone building in the center of town, Internet banks must offer trust in different ways. For established banks, the task would be as easy as "moving" the trust from the physical to the digital bank; for others it is not as easy.

However, many states operate with a bank guarantee, which means that if the bank fails, a guarantee fund will secure your savings. In many ways this seems to be the simplest way of getting customers to trust an online bank. Confidence in the government guarantee does the job. This guarantee clearly has an impact in terms of promoting new bank services.

Customers also need to trust the technology and their own ability to master the devices and the user interface of the bank. Initially, many systems had serious drawbacks. For example, there are cases of customers' typing errors causing large amounts to go to the wrong accounts. There was a famous case of this in Norway where a customer lost nearly US$50,000 just by adding an additional digit to an account number. The problem was that the bank did not maintain any log information, so it was impossible to know if the customer had typed in an additional digit or just typed the incorrect account number. In the former case, the fault was clearly with the user interface; in the latter, the customer was to blame. By simulating the user interface and asking a set of volunteers to enter 2000 transactions, it could be proven that the customer had entered an additional digit. What should have resulted in an error message from the user interface ended up as an erroneous transaction. In the end, the bank backed down and offered complete reimbursement.[2]

Modern interfaces have improved and users have become more experienced. Many of the latest smartphone interfaces are also excellent. It often seems easier to develop a new good interface for new technology than to improve an existing interface. The advantage in the first case is that there is no previous history, either user experience or technologically, that must be taken into account.

Login and Security

Many systems are used for logging into an online bank account. Some provide us with a calculator that offers one-time codes, others with a card that has a set of codes. The one-time codes are to be used together with a PIN. Today, most users rely on their smartphones. The online systems offer one or two keywords on the PC screen and then send a message to the smartphone with the same keywords to confirm contact between the Internet site and the phone. When the user has confirmed that the keywords on the phone are similar to those on the screen and typed in the PIN code on the smartphone, the login is executed. In some countries such a system is used for all secure logins, both to the Web sites of private companies, such as banks, and to the more official, such as the tax office. When the same system and the same passwords are used, login becomes much simpler, especially when the systems that we use less frequently have a similar login procedure to the systems that we use frequently.

These forms of two-pass security are fairly secure. But if malware has infected both the phone and the user's PC, it is possible to break in to Internet bank accounts.

In a practical world, there needs to be a balance between security and convenience. Perhaps customers should be allowed to set their own standards—that is, banks should allow for customized security in the same way as a browser allows us to set the security level. For example, we can set up a browser to disallow scripts, cookies, or Java applets. Similarly, maybe we should be allowed to specify that our bank account should be accessed only from our smartphone, the office, or home computer, or that we want to provide an additional password for transfers of large amounts or for the use of an ATM.

Customized security recognizes that customers are different. While some may access their bank accounts from many computers, others use only a smartphone. Some customers may have small amounts in their accounts, while others may have larger amounts available. Of course, the bank has to require a minimum security level, but if we can easily reduce the risk of an attack, and the hassle that follows, why shouldn't we have this option of customizing our own risk management? I shall return to this discussion in Chapter 15.

Formalized Data and Functions

Banking data and bank transactions are formalized and symbolic, which makes them ideal for being handled by the computer. Since data are compact, a lot of information can be offered on a single display line; the computer screen may then provide an overview that is difficult to get using telephone banking, perhaps even better than if one is physically present at the bank. Another advantage is that the computer will present the updated picture, not the one from last week as is the case with the printed version that is sent by mail.

Internet banking may be offered as an alternative service to traditional banking, or by new "true" Internet banks that base all their contact with customers on the net. These banks try to offer a full suite of functions, similar to traditional banks. Personal contact with customers is through email, chat systems, or by telephone; some are also starting to use video calls. Paper is used only when required by law (for example, for the initial account setup contract). Customers access their accounts directly, using the Web as an interface to the banking system. They perform their own transactions, check balances, pay bills, print account statements, and so on.

Since bank functions are highly formalized to a large degree, good user interfaces will enable most people to perform them with ease. The data needed is minimal, such as amounts and account numbers, which implies limited keying. As these services evolve, the need to key in data, for example from invoices, will be greatly reduced. Regular invoices, such as those from utility companies, can be entered for automatic payment, where the bill will be paid on the due date if the system is not told otherwise. Other requests for payment can be sent to the customer as an email message and as an electronic record to the banking system. When the customer accesses the banking system to pay the invoice, all data will be available so that the invoice can be paid just by a confirmation—for example, by checking off the box next to the invoice.

All these functions are formalized and quite easy to implement, but modern banks are now also in the process of automating loan processes. Instead of having to wait weeks for the approval of a loan, many banks can now give an answer within minutes as long as the necessary background data is available in digital form. For example, to approve a mortgage the bank will need information on income, the customer's credit rating, and the property that is to be bought. In countries where information is available in digital form for the loan processing algorithms, the process can be automated.

The Advantage of Being on the Internet

An Internet bank does not need branch offices, human tellers, cash or check handling, or printing and mailing of statements. In fact, the daily operations of an Internet bank are in many ways just a computer, a server. The customers themselves have replaced the tellers; the branch offices are replaced by the customer's home and office, or by the customer's smartphone. Customers use their own hardware and software to access the bank system and they pay their own communication fees. Since the customers themselves handle the manual part, all other services can be performed automatically by the banking software. Thus, Internet banks should compete well both on fees and on interest rates.

There are also additional advantages for the customers. They can access their accounts 24 hours a day, seven days a week, 365 days a year. The bank's website is available from everywhere, even mobile phones. If all transactions are electronic—that is, with no checks or other paper in the system—the account balance will always be updated. All bills entered into the system will be paid on their due date, and the system can provide an overview of upcoming payments. An additional advantage of removing paper from the system is that the customer no longer has to archive manual transaction records. Instead, all historical data are available from the Web interface, in the customary chronological presentation or, for example, presented by the addressee. With one click, the customer can get a list of all transactions regarding the credit card company or any other institution or account. Previously, customers had to do this themselves by going through a large number of printed bank statements.

Personal Service

While a pure Internet bank is just a computer, in practice some personal services are needed to handle special cases. Examples include when the customer has technical problems or does not understand the interface, or when the interface does not provide the functionality that the user requires.

Pure Internet banks try to handle these exceptions by providing customer service through telephone, chat, and email. But this affects costs and therefore also the interest rates and fees that the bank can offer. The business model really requires that personal service be offered only for exceptions. Therefore, it is important that the bank monitors all personal requests to see which may be handled by an improved or extended user interface. In this way, similar requests may be avoided in the future.

This is not always possible. In principle, Internet banks use a cash-free model in which all transactions are completed electronically. However, even though digital payments are replacing paper money for many transactions, some customers still use cash. Cash withdrawals are handled by cash-back services when using the plastic card in a store, or by ATM machines. However, pure Internet banks without physical offices cannot easily offer the opposite service of depositing large amounts of cash into an account. These banks may not be a good alternative for customers who are paid in cash today, such as newsstand agents in countries where cash is still in use.

A Good Bank Infrastructure Is Important

Many countries have a bank infrastructure that makes it easy to establish Internet banks. In Scandinavia, which has centralized bank clearing systems, nearly all bills are paid online. Costs are an important issue here. The average transaction cost for paying a bill on the Internet is approximately US$0.03, compared with an average of $0.60 for paper-initiated transactions. With even higher transaction volumes, Internet-based solutions will be even more cost-effective; therefore, banks often offer free services.

In this respect, small countries may have an advantage since it is easier to agree on and establish national systems and standards for account numbers and bank-to-bank transactions, for example. Such a formalized infrastructure, which in many cases was established long before the Internet, can provide the foundation for an efficient all-electronic system, independent of cash and checks. With such a system, all deposits are made electronically as bank-to-bank transfers. Bills can be paid the same way, by an electronic transfer directly to the recipient's account—independent of his or her choice of bank. Start-up online banks may also find it easier to establish trust in small, homogeneous countries. As we have seen, a state guarantee for protecting the funds in the bank also comes in handy.

It is interesting to see that while some countries are well into the digital economy, others are lagging behind. Iceland and Norway are examples of the former. Here, nearly all transactions are electronic, Internet banking is the norm, all businesses accept card payments, and banks are closing their branch offices or turning them into cash-free offices. The United States can be used as an example of the latter. Here, checks are still common and many banks have expanded the number of branch offices.

However, while a transaction costs just a few cents in Norway on average, it is much higher in the US. One usually looks at Germany as a technologically oriented country, but here cash is used for 80 percent of all transactions, compared to less than 3 percent in the Scandinavian countries. We see that the "Internet countries" get a clear strategic advantage, similar to that gained by those countries that were among the first to develop good railway systems in the nineteenth century.

However, there is no doubt that the Internet will soon become the primary medium for bank transactions in most countries. The advantages of online banking are so great that they will remove nearly any barrier.

Conclusion

For ordinary customers, there is no longer much need to go to the bank. The bank can be accessed from a home computer or a smartphone. Many of the processes, such as paying utility bills, can be set up to be handled automatically, requiring no involvement from the customer. Simple loan applications can be handled by the computer. When the computer handles all the high-volume standard processes automatically, there will be ample room for offering personal service to the customers that need this and are willing to pay for premium service.

In Chapter 15 I shall speculate about what the banks can become in the future. For example, can smart software in the form of robots become our personal economic assistants?

Notes

[1] Parts of this chapter are taken from Olsen, Kai A. (2012) *How Information Technology is Conquering the World.*

[2] The case is described in detail in Olsen, Kai A. (2012) *How Information Technology is Conquering the World: Workplace, Private Life, and Society,* Scarecrow Press, December 2012, Lanham, Maryland, Toronto, Oxford, ISBN 978-0-8108-8720-6 (paperback) and 978-0-8108-8721-3 (e-book).

Chapter 11
Virtual Currencies

A currency can be represented in many ways: as coins, banknotes, or as bits in a computer. Most of the currencies that are common today are available in a physical form, but some do not have a physical representation. We call these virtual currencies.[1]

Examples of virtual currencies include frequent flier miles, bonus points, currencies that are used within computer games, and general currencies such as bitcoin and ethers (the currency based on the Ethereum platform).[2] Both bitcoin and ethers are cryptocurrencies that use blockchain technology, a digital "ledger" that is universally acceptable. New transactions are bundled together in a block and added to the chain. This is performed using cryptographic techniques so that it is easy to verify the transactions, but practically impossible to alter them. The idea is that everything is open, nobody is in charge, and anybody can verify the transactions.

Several financial firms are experimenting with blockchain technology—for example to register trades. While the blockchains used for bitcoin and other cryptocurrencies are open, the blockchains used in the financial industry are only accessible for customers who identify themselves through digital keys. In many ways, these "permissioned" networks implement the exact opposite of the ideas behind bitcoin.

Some of these digital currencies can only be used within a closed community and cannot be transferred to other currencies. An example is initial coin offerings (ICOs), where companies offer their own digital "money." Later on these tokens can be replaced by real money or by other services. However, in July 2017 the US Securities and Exchange Commission issued a warning that ICOs may be determined as securities, and would therefore be "required to comply with federal securities laws, regardless of whether those securities are purchased with virtual currencies or distributed with blockchain technology."[3]

Money used within a game is another category; it may be bought with real money but can only be used within the game. Frequent flyer miles can be used to buy tickets, upgrade a ticket, pay for items on a flight, pay for car rental and hotels, or shop on an online store. Other virtual currencies, such as bitcoin, can be bought and exchanged into traditional currencies. There are even ATMs that will sell you bitcoin.

Most of these virtual currencies exist as private currencies in an unregulated form, but the issue may be more complicated when some of the central banks are planning to create their own virtual currencies.

Virtual National Currencies

A few central banks are discussing the possibility of issuing virtual banknotes; one of these is the Riksbank in Sweden. Cecilia Skingsley, deputy governor at the Riksbank, said, "Will we have e-krona in an e-wallet in the future, as naturally as we now have a wallet with cash in it? The less those of us living in Sweden use banknotes and coins, the clearer it becomes that the Riksbank needs to investigate whether we should issue electronic money as a complement to the money we have today."[4] Sweden was the first country to introduce banknotes (in the 1660s) and may be one of the first to introduce a digital currency. Skingsley continued: "The declining use of cash in Sweden means that this is more of a burning issue for us than for most other central banks. Although it may appear simple at first glance to issue e-krona, this is something entirely new for a central bank and there is no precedent to follow."

Japan is now following. It plans to have launched its own digital currency—called J Coin—by 2020, to counter the mobile pay systems that are introduced by the Chinese company Alibaba. J Coin will be pegged to the Japanese yen on a one-to-one exchange rate. The Bank of England has its own experiment called RSCoin, developed in partnership with researchers at University College, London. RSCoin uses a different protocol to that of bitcoin, with the aim of increasing the number of transactions that can be handled every second. Bitcoin currently has a cap of seven transactions per second, while Visa, for example, is able to process at least a thousand times as many.[5]

Virtual currency is not intended to replace cash, but to be an alternative to cash. It can be implemented by the use of blockchain technology. The idea is that it can be used as cash, where transactions are performed immediately. The central bank will issue the digital notes and will also guarantee them, just like cash. In contrast to the public currencies such as bitcoin, the currencies offered by the central banks will be closed systems where only trusted participants can access the ledger.

Technically, such a currency can be made traceable or anonymous. From the discussion it seems that this is viewed as an alternative to cash and that the notes will therefore be made as anonymous as cash. At the same time central banks will, of course, have to follow all the regulatory requirements for the financial industry. An advantage is that credit card companies and banks are taken out of the loop, thus reducing the amount of fees.[6]

Does this imply that the central bank will compete with ordinary banks? Perhaps customers can have an account here, instead of in ordinary banks? In his book entitled "The Curse of Cash," Kenneth Rogoff offers a warning, arguing that these virtual banknotes will be used for criminal activity and to avoid paying tax. In this respect, the virtual banknotes can be viewed as an alternative to the cash cards that are in use today.[7]

In a modern society it is important to collect tax, maintain regulations in the workplace and elsewhere, and stop international criminal activity. In order to make life more difficult for criminals, hinder terrorists, avoid an underground economy, and finance schools, health care, and many other important government functions, it is important to have a good overview of all payments. The digital systems we have today provide such an overview, while cash and these anonymous virtual banknotes do not. Therefore, we should look forward to cash disappearing, not introduce new tools for criminals.

There is no technical reason why virtual currencies offered by central banks should not offer the same possibilities for tracing as other virtual currencies. This will offer the best of both worlds. A low-cost digital currency can be trusted since it is provided by a central bank, while also offering the other advantages of digital payments, thereby ensuring that the currency is not used by criminals, that taxes will be paid, and so on. However, it is too early to say how these currencies will be implemented.

Central banks have the privilege of printing money and this is an important source of income for these banks. As customers move away from cash, this income—known as seigniorage—is reduced. The idea of creating an official virtual currency may be an effort to maintain some of this income. I shall discuss these issues in more detail in later chapters.

Blockchain and Bitcoin

The success of a new currency is based on

- Limiting the amount of money created to avoid hyperinflation and a valueless currency
- Offering functionality to store money and use money for transactions
- Avoiding counterfeiting.

Of course, it is also important to create interest around the currency. The cryptocurrency bitcoin managed to fulfill these requirements. Each transaction is registered in a distributed ledger, called a blockchain. Using cryptographic techniques, the blockchain is protected by coding the last transaction into the chain. This is a revolutionary approach. There is no "center" that one has to trust. Blockchain technology ensures that the transactions are registered in a tamper-proof way. Thus a group of strangers can work together to maintain the currency. There is no need to trust the participants. The cryptography techniques offer proof of concept.

In addition to details on the transaction, moving a bitcoin from one account to another, a number is added to the blockchain. This number is not easy to find. Using special-purpose computers, "miners" use algorithms to try different numbers. The idea is to find a number that fulfills a set of requirements. If successful, the miners are offered a transaction fee and new bitcoins. This is the reward for performing these "banking" functions. The average numbers that miners have to test is more than 200 quintillion—that is 10^{18} or 10 multiplied by itself 18 times. The complexity is automatically adjusted to the number of bitcoin "mined," thus regulating the amounts of bitcoin that go into the market. All in all, the software will ensure that a maximum of 21 million bitcoin can be generated.

A negative side of the mining is that it requires a lot of electrical power—as much as is used by all of Denmark or Ireland! One argument for an online economy is that it consumes less energy than when paper money and coins have to be moved by truck. However, bitcoin will use more energy than the incumbent currencies, at least in digital form. While a credit card transaction can be performed with a minimum of energy, the bitcoin servers consume a lot. Today a large part of the mining is performed in Inner Mongolia, where power is cheap. But since most of it comes from coal-fired plants, it is not environmental friendly.

There are more than 1400 other cryptocurrencies today, but bitcoin is the most famous. It was introduced under the name of Satoshi Nakamoto in 2008 as open-source software. Since then there has been a lot of speculation about who Nakamoto is, and this speculation may have added additional interest to bitcoin.

Bitcoin is a distributed currency. There is no "central bank"; in fact, there is no "center" at all. After a miner has successfully added a transaction to the blockchain, this is broadcasted to other nodes. While it is difficult to add a transaction to the blockchain, verifying its correctness is easy. A new block is created every ten minutes. This consists of a group of accepted transactions and is published to all nodes. This is done to avoid a bitcoin being used twice, but also implies that one has to wait for the block to be distributed before being sure that the transaction is recognized.

The advantage of bitcoin is that it offers anonymous transactions, but perhaps only to a certain degree. Combined with other data, such as IP addresses, it may be possible to identify stakeholders. The Norwegian Police, for example, explained that "blockchain technology makes all transactions public. Combined with other investigation material we have methods to catch criminals that use cryptocurrencies."[8] There are other cryptocurrencies, such as zCash and Monero, that offer a greater degree of anonymity and are therefore more interesting to criminals. However, all legal exchanges must also follow regulations such as "know-your-customer" (KYC) verification to anti-money laundering (AML) procedures.[9]

For tax authorities it may not be so easy. While it is possible to identify the stakeholders behind a bitcoin transaction, this will require significant resources. So while traditional digital payments always have an identified sender and receiver, offering the possibility of following transactions automatically, this will not be the case for bitcoin.

Bitcoin can be used for person-to-person transactions. The transaction fees are voluntary, but one may have to pay to get the transaction into a blockchain and block within reasonable time—that is, the miners will put a priority on transactions that pay. Today the cost of a transaction is between $4 and $50 and it can take from 10 minutes to several days to confirm.[10] Thus, bitcoin has been more a currency for speculation and representing wealth than a currency for making payments. In fact, we may call it a failure as a currency as it fails on all requirements—it is too volatile for pricing, too slow and expensive for monetary transactions, and too volatile for storing value. If your pension fund was in bitcoin it could double one day or be halved the other.

In order to improve the use of bitcoins for transactions, Bitcoin Cash was introduced on August 1, 2017. The cash variant inherited the transaction history up to that date; after this "fork," transactions were separate.

A disadvantage of bitcoin and any other "private" currency is that they exist at the mercy of the central banks. At any time, for example, in order to stop crime or terrorism, the central bank can interfere, perhaps by declaring the currency illegal or taxing its use. The major advantage of using traditional currencies is that these are backed by the central banks. Central banks have their own interest in protecting the currency and will work hard to keep it stable. There may be a financial crisis, the currency may lose value compared to others, but a central bank will never be the victim of a bank run. As we shall see, bitcoin as a "private" currency is vulnerable.

The value of bitcoin with regard to other currencies has been very volatile. It was originally valued as 0.30 to the US dollar. In 2017 its value ranged between US$1100 and nearly US$20,000. The value of other cryptocurrencies has also soared, also after a drop caused by restrictions by Chinese authorities, and later depreciated.

While some people view bitcoin as the new gold, others offer warnings of a pyramid scheme.

Over 100,000 merchants accept bitcoin; the advantage for them was originally that the fees were lower than for credit cards, but this is no longer the case. Several nonprofit organizations accept bitcoin, as do many companies that sell entertainment on the net. Still, only a few retailers accept bitcoin.

Bitcoin and other cryptocurrencies have been used for buying narcotics, child pornography, and weapons. Clearly, the stakeholders value the apparent anonymity of the currency. The Silk Road marketplace, a platform for selling illegal drugs, relied on its customers using bitcoin. It was closed down by the FBI in 2013. In May 2017 a worldwide cyberattack, the WannaCry ransomware cryptoworm, was launched that infected more than 200,000 computers. The plot was to make data unavailable by running it through a cryptographic process, demanding a bitcoin ransom for the decryption key. Other hackers have tried to blackmail Disney[11] and Netflix,[12] threatening to release early versions of movies and series if ransoms were not paid. Payments in anonymous bitcoin were an important part of all of these schemes.

There have been many reports of bitcoin being stolen. Since the transactions are anonymous and not possible to validate, hackers may be able to steal passwords or alter the receiver before a transaction is performed. If a password is lost, the bitcoin will also be lost. There have been many black market sites that steal bitcoin from customers—for example, by not providing the services that users have paid for.

While the decentralized nature of bitcoin offers advantages for some users, there is no central bank to stabilize the currency and little security for consumers. Many exchanges for cryptocurrencies have been hacked and users have lost large amounts of money.[13] While a credit card company will refund money if the goods it has paid for are not delivered, this does not apply with bitcoin. Bitcoin will already have to compete with new services that are based on traditional currencies, such as simple person-to-person transactions that offer immediate payment based on a smartphone. These payment systems are already offered by many banks, and Facebook also provides a friend-to-friend payment service through its Messenger system.

Many people see the blockchain as a promising way of ensuring the integrity of any type of transaction. Thus, it may have an interesting future apart from being the technology behind bitcoin and other cryptocurrencies.

Reward Points—a Currency?

Where is your money? You may answer that some is in your wallet, most is in the bank, and that there may be a few banknotes and coins in the car or at home. However, most of us have something else: frequent flier miles, discount coupons, reward points for shopping, game credits, and more. In some sense we may look at this as "money," but it often has limited use. Game credits can only be used on one computer game. Frequent flier points can be used to pay for tickets, but only within a particular airline or alliance and their business partners. Some reward systems are more open and allow customers to convert their rewards to an ordinary currency—for example by transferring to a bank account. Others offer the reward as an immediate discount on the final amount.

The reward system raises many questions with regard to income tax. Business travelers may receive "miles" or points in their own name that can be used for private travel. Should they be taxed for this "income"? Is it a misappropriation of funds that belongs to the company if the points are used to buy private tickets? This also raises the question of the value of a point. This is not so easy to compute. A traveler may use 5000 frequent flier miles to save $100 on a ticket or $200 on another.

Today there are many opportunities for creating these currencies. They can be protected as ordinary currency or by using a blockchain, as explained in the previous section. IT has made the cost of maintaining these systems acceptable and companies see it as an important way of retaining loyal customers. The importance of these "currencies" has been limited to date, since they have to be used within the company. This is perhaps also why authorities have allowed these "currencies" to exist and have avoided enforcing a strict interpretation of tax rules in most countries.

Some years ago Facebook offered a virtual currency called Facebook Credits. It was expected that this would be developed into a micropayment system available to all Facebook applications. However, Facebook has announced that it will discontinue the service and instead rely on local currency. One reason may be that the standard digital payment systems are also able to handle small payments, and also that the often unlimited subscription model seems more interesting for customers than having to pay for each file, whether it is a piece of music or an article. Facebook may also have seen that it would run into difficult legal territory if its credit system became a currency.

New Providers

Fully automated payment systems will make it quite easy for new providers to enter the scene using the standard currencies. Apple Pay is an example; this digital wallet service offers payments through an iPhone, Apple Watch, iPad, or Mac. Facebook offers a friend-to-friend payment service, while Google has its "wallet." In China, both Alibaba and Tencent have their own payment systems, and Xiamo, one of China's largest smartphone brands, is joining in with an interest-bearing wallet. Ant Financial, a payment company affiliated with Alibaba, is very popular in China; it is 16 times larger than PayPal and has ambitions of becoming more international.

The social media system WeChat (or Weixin), developed by Tencent, has been very successful in China. It offers a payment system that enables users to pay bills, order goods on the net, transfer money to other users, and pay in stores. It also includes a wallet where users can store money. While users have offered bank account information to WeChat, the system is also a bank in itself.

This may pose a threat to traditional banks and credit card companies. These new services will be online only and will not have the costs of handling cash. A clear advantage is that the payment systems are integrated with other functionality, as instant messaging, a social network, online shopping, video broadcasting, games, and so on. Here we have a similar situation to the one discussed in Chapter 9 on commuter tickets, where buying the ticket was just a part of a larger transaction; that of finding the right transportation route, getting the schedule, and so on. With WeChat and similar systems, the actual payment is also an integrated part of something bigger, such as sending money along with a greeting. There is a custom in China of exchanging money between friends and family members during holidays in the form of red envelopes. WeChat captured this tradition when it implemented a similar "red envelope" function in the system—that is, a system that could take care of the complete idea, both a greeting and a money transfer.

While there is currently not much revenue, and very little profit, to be made from running a payment system, the advantages are that one may get valuable data. Still, the problem may be that only part of the picture is obtained; for example, a bank may have data on the payment, but not on the context. Clearly, it would be very valuable if it was possible to get users to perform all tasks on one platform, such as with WeChat. Companies could then perform extensive data mining to get a more complete profile of customers. If this is done immediately, the customer can be "captured" while performing a transaction and offered relevant information. For example, instead of "other customers bought," a company could offer "you may also be interested in." With a complete user profile, there would be no limit to what a retailer could suggest. In practice it would be possible to implement an automatic personal assistant.

A personal assistant can push information that is relevant to the user and can initiate actions that it finds necessary. For example, a system that has all the data on utilities may suggest a change of provider. In the next step, the customer may allow the system to perform the change as long as it results in a reduced bill. This will allow the system to ask for offers from utility companies for its entire customer base. In order for the assistant to do its part, it needs access to all data that is relevant for the user. I shall return to these issues in Chapter 15.

Conclusion

Technology has been used to allow traditional currencies to be represented in a digital form. This opens the way for Internet banks and for making digital payments. However, the technology also allows for the creation of new virtual currencies that only exist in a digital form; bitcoin is a good example. In the long run, traditional currencies may also end up as true digital currencies as one get rid of cash.

The main difference is that some currencies are supported by central banks, in principle by governments. In most cases, this ensures a stable and safe currency. The other, "private" currencies do not have this privilege, which poses a risk. For example, one could speculate about what would happen if bitcoin or any other private currency changes from interesting experiments into a competitor for euros and US dollars. Would they be allowed to undermine the traditional currencies or would the central banks set up roadblocks, ranging from taxation to making these currencies illegal? Private currencies may be tolerated as long as the amounts are small or where one has strong restrictions for their use, but may be stopped if they become an alternative to ordinary currencies.

Notes

[1] The term digital currency has also been used, but since all currencies may have a digital representation I shall continue to use the term virtual currencies.

[2] Both Bitcoin and Ethereum are blockchain-based currencies.

[3] Investor Bulletin: Initial Coin Offerings, https://www.sec.gov/oiea/investor-alerts-and-bulletins/ib_coinofferings

[4] http://www.riksbank.se/en/Press-and-published/Speeches/2016/Skingsley-Should-the-Riksbank-issue-e-krona/

[5] Zitter, L. (2016) "The Bank of England's RSCoin: An Experiment for Central Banks or a Bitcoin Alternative?" *BitCoin Magazine*, March 28.

[6] Evans-Pritchard, A. (2016) "Central Banks Beat Bitcoin at Own Game with Rival Supercurrency," *The Telegraph*, March 13.

[7] Kenneth Rogoff (2016) *The Curse of Cash*, Princeton University Press.

[8] http://e24.no/lov-og-rett/teknologi/politiet-har-knekt-bitcoin-koden-kriminelle-vender-tilbake-til-kontanter/24166982

[9] See a global legal and regulatory guide to cryptocurrencies here: http://www.nortonrosefulbright.com/knowledge/publications/139847/a-global-legal-and-regulatory-guide-to-cryptocurrencies-chapter-6

[10] *The Economist*, June 3, 2017.

[11] http://www.telegraph.co.uk/films/2017/05/16/hackers-demand-ransom-disney-pirates-caribbean-5/

[12] https://www.theguardian.com/media/2017/apr/29/hacker-holds-netflix-to-ransom-over-new-season-of-orange-is-the-new-black

[13] https://www.theguardian.com/technology/2017/dec/07/bitcoin-64m-cryptocurrency-stolen-hack-attack-marketplace-nicehash-passwords

Chapter 12
Advantages of a Digital Payment System

Digitalization is a strong force. Most jobs that are formalized, where all or parts of the processes can be described as a computer program, have been affected either directly or indirectly. In industry, the use of robots is increasing. We now shop, buy tickets, and communicate online. Digitalization is especially beneficial when there are a large number of transactions, as is the case with all forms of payments. In many ways, this is an ideal application for the computer. Data is initiated in a digital form, from a cash register or a website. It is compact and formalized. With universal access to computer networks, at least in developed countries, transactions can be performed immediately from anywhere. In addition to using fixed terminals or websites, just about everything can be done with a smartphone. For many types of payments, such as person-to-person or on the web, smartphones have many advantages. In underdeveloped countries, the mobile phone may be a good solution for operating in the digital world.

In this chapter, with the understanding that there will be a choice of many different forms of online payment systems, I shall discuss the advantages of a digital economy. Clearly, some of the advantages, such as the possibility of tracing digital transactions, may also be seen as disadvantages because they may reduce privacy. This chapter will look at the positive side of a digital economy, before the next chapter offers the opposite view. Note that in both of these chapters we discuss traditional digital payments—that is, where one pays with a credit card, a debit card, or a mobile phone. All of these payment systems are connected to accounts owned by the user, which means they are traceable. As we have seen, there are different digital payment systems, from cash cards to cryptocurrencies, that offer a greater degree of anonymity.

Fighting Crime

The anonymity of cash is attractive to criminals. In the workplace, cash is a serious problem because businesses that have income in cash can avoid paying value-added tax and income tax. Further, they can avoid fulfilling work and pay regulations. This allows for "social dumping," which involves hiring people who do not have a work permit and forcing them to accept very low pay. In this respect it becomes very difficult to regulate the workplace; the authorities lose tax income and the workers are not a part of social security or pension schemes. Perhaps even worse is the fact that these black market operations may undercut legal businesses. By evading tax and paying their workers less than the minimum wage, they get a competitive advantage. This is especially the case in construction, agriculture, and hospitality businesses.[1]

In order to maintain a social security system, it is important to have a good overview of the income of the applicants. A problem may be that income in cash is not reported, which creates the risk that people with a reasonable income will also receive unemployment benefits. People may avoid paying taxes at the same time as they receive benefits that they are not entitled to. Not only does this make it more difficult to maintain the economy of a social security system, but the system may be discredited in the general population.

Anonymous cash is an ideal tool when the idea is to hide wealth from authorities—for example, to avoid paying taxes or to hide illegal income—since it is so difficult to control. A cash economy is important for all types of criminals. If you are mugged on the street, the mugger can use your cash as if it was his own. A thief may break into your house and sell the items he steals for cash. Narcotics dealers also rely on the anonymity of cash.

Cash is bait for criminals. Where do we find this bait? In stores, gas stations, in taxis and buses, and many other places. The famous bank robber Willie Sutton was once asked why he robbed banks, to which he is said to have replied, "Because that's where the money is."

Employees handling cash are in the danger zone and may risk injury. Even if actual cases are low in most countries, the risk is a stress factor. In one survey, 20 percent of hotel and restaurant workers in Norway said that they have felt insecure at work due to the handling of cash.[2] Unions for bus drivers are demanding cash-free payment systems. In countries where most of the economy is digital, the few places in which cash is still used become targets for criminals.

In most digital-based fraud, such as where criminals manage to get illegal access to bank accounts, the transfer from digital money to cash is an important ingredient. Clearly, retaining the embezzled funds in a bank account is risky. To avoid the problem of having the money confiscated, it is immediately withdrawn in cash. Without this possibility, many fraud schemes would be impractical. Of course, a transfer to any other anonymous currency, such as a cryptocurrency, would cause the same problem.

While it would be naïve to believe that crime will disappear in a cash-free society, the life for criminals would certainly be much more problematic. As we have seen, money in a form that is generally accepted in a society boosts commerce. It is also very convenient. Without access to money, criminals have to return to pre-money times, for example by using precious metals. The drawbacks are that their customers will have limited access to gold and silver, sellers will need competence to verify the metals, and the transaction will therefore be much more complex. In a cash-free society, criminals can also move to the digital economy, but this may not be easy. Unlike cash, digital payments in traditional currencies are not anonymous. Even if the sale of narcotics is camouflaged as legal sales, there is always the risk that authorities may detect the fraud and then follow up on earlier payments. The risk for the criminals and their customers will be much greater than when using cash.

Nobody will rob a bank that doesn't have money. Without any chance of finding cash, criminals will look for other targets. Retailers selling expensive items like jewelry, watches, liquor, and cigarettes that are easy to sell may be at risk. Ordinary citizens may be robbed of everything from outboard engines to cars and electronics. This is not new, but when cash is no longer a possible target, the criminals will find themselves in a more difficult situation than before. It may be not be easy to dispose of the stolen goods within a cash-free society. Their customers for the stolen goods will not be able to pay in cash. They could return to a barter system, but that is inconvenient. Another possibility is that the criminals may establish some sort of underground currency. However, the impracticability of such a solution would be clear if there was no way to convert it into the ordinary currency.

Another, more realistic option would be to sell the goods abroad, in another so-ciety that still uses cash. This is sometimes the case today. As long as there are coun-tries that use cash, these may be havens for criminals, at least until those countries also go digital.

In the fight against terrorism, one weapon is to control their funding. This is much simpler to do in a digital economy. While transactions may be hidden as some-thing legitimate, there will be more information about who is involved. The online transaction will be an important factor in getting insight into a terrorist network.

Limiting the Black-Market Economy

Modern societies are dependent on tax income to finance schools, health care, roads, defense, and much more. The tax that is most relevant for payment systems is the value-added tax, a percentage based on the price of each item. As we have seen, cash involves possibilities of hiding income and participating in the black market economy.

To control cash, authorities usually demand that the cash transaction is regis-tered in a cash register. However, there are many ways of avoiding paying tax on the full amount. Payments can be registered with a lower total or not registered at all. There are ways to supervise payments in cash, but these controls are manual and therefore time-consuming and expensive. Unofficial income in cash can be used to hide profits and to pay employees off the record, thereby also cheating on income tax. In addition, these unofficial payments of salary make it possible to avoid following rules regarding minimum wages, pensions, and health insurance.

In the same way, cash can be used to pay plumbers, electricians, and many other service providers unofficially. This black market economy may be convenient for customers and advantageous for the workers who can avoid tax, but again the society as a whole will miss out on tax revenues. Such cheating can lead to unfair competition; the company that avoids paying tax can offer lower prices than those that have to include taxes.

In an online economy, it is simpler to control that everybody pays taxes. It is much more difficult to hide online transactions than cash. In a digital society, where all the background data is available in digital form, tax authorities can easily control that every income is included in the accounts. This is so simple that nobody will try to cheat as one could do with cash, just avoiding registering a sale. However, authorities can also use advanced software such as data mining techniques to pick out businesses for a more extensive control—for example, to find the restaurants that seem to have lower revenues than comparable establishments. These can then be candidates for a more extensive control.

In a cash-based system, the tax office may check whether the amount on a re-ceipt is also found in the ledger, but this is impractical and will take time. There may also be ways of cheating, such as only avoiding registering the transactions where the customer did not demand a receipt, that are difficult to detect. In a digital economy the comparison of payments with ledgers can be performed automatically. For exam-ple, if all civil servants pay with a credit card, all information on expenses will be part of their electronic travel and subsistence claim form. All these transactions can then be compared electronically with the ledgers of the different merchants. As a next step, one may also include travel data from the private sector. In such a system it will be nearly impossible for taxi drivers, restaurants, hotels, and so on to avoid paying tax.

Difficult to Hide Funds

It will be difficult to hide funds in a cash-free economy, even in an economy where only the high denomination notes are removed. With no cash, or only small denomination notes, it will be extremely impractical to handle large amounts in cash. A million dollars in $100 bills will weigh 10 kg (22 pounds) and the height of the stack will be 1.24 meters (48 inches). If the highest denomination is $10, the above values can be multiplied by 10, or by 100 if $1 bills are the highest.

Funds can be hidden abroad. This was the traditional way to avoid tax, but has become much more difficult in later years. Traditionally, banks have protected customer data, and this was an important part of the business model of many international banks. Today, even the Swiss and the Liechtenstein banks that were renowned for protecting their customers have been breached by disloyal employees. The German government has been more than willing to buy these data and has paid millions of euros for access. But this is just a fraction of the money that Germany has been able to retrieve from tax evaders. Germany has also shared these data with other countries. We see that IT systems have made it easy to steal large amounts of data, exemplified by these cases and, not least, the case of Edward Snowden, who collected huge amounts of data from the NSA (National Security Agency in the United States). What was impossible in a paper-based society is today quite easy, at least for people who have access.

Today most countries share data on bank accounts. Pressure on the few tax havens that try to hide data is increasing, and most have been obliging in this regard. In fact, the breaches of security in later years have shown that it is very difficult to hide information, especially when one's own employees are involved, as was the case with the Swiss banks. The Panama Papers case, when 11.5 million documents were stolen from the Panamanian law firm Mossack Fonseca and leaked to the press, caused embarrassment for many well-known people. In many ways the situation here is similar to that of digital music. Pirates pressed the music business to open the door for digital services; here, the leakage of information forced institutions around the world to share data on international customers.

Those who want to hide funds can buy gold, precious stones, or art; however, since payments have to be done online in a cash-free society, there will be no way of hiding the transactions. In practice, it will be very difficult to hide funds.

Reduced Costs

Handling cash is expensive. Notes needs to be printed and coins minted and it needs physical transportation. Since cash is anonymous, there must be secure routines in all phases, counting the cash, checking amounts against what have been registered, and more. With cash as bait for criminals and disloyal employees, safe storage mechanisms are needed, such as cash registers, safes and surveillance systems. Even in a small country like Norway, with five million inhabitants and a flourishing digital economy, the cost of handling cash is estimated at close to four billion Norwegian kroner (approximately US$470 million[3]) every year.

Nearly all of these operations are unnecessary in a digital system. There will be no counting; money and transactions are sent over computer networks, and all balancing of accounts is performed automatically. There is no need for paper, safes, or surveillance systems. There will be other costs, of course, which are discussed in Chapter 13.

In addition to these direct costs, there are indirect costs. People who handle cash may risk being physically hurt in a robbery or may feel insecure when handling large amounts. There may also be problems when the contents of the cash register do not match what is registered. This may be due to an error, but most employees involved will be uncomfortable in such a situation. Some years ago, a survey of Norwegian restaurant employees found that guests who paid with cash gave larger tips than those who paid digitally.[4] Despite this, 86 percent of waiters said that they preferred that guests paid digitally. Clearly, the advantage of not having to handle cash was found to be greater than the extra tip.

Better Data

The times we live in have been called the fourth Industrial Revolution. The first Industrial Revolution began in the late eighteenth century, when machines replaced manual labor. The next came when Henry Ford introduced mass production, making industrial goods widely available.[5] In the third, computer technology was used to automate production. While these previous revolutions have been fuelled by coal, oil and electricity, data is seen as the main ingredient in the fourth Industrial Revolution.

When all devices are on the Internet—everything from personal health monitors to cars—enormous amounts of data are generated. This opens the door for new interesting applications. For example, with access to online data on other cars, a navigation app may give you updated traffic information and use this to suggest the best route. Taxi dispatcher apps, such as the one from Uber, can find the nearest available car and also predict when it will arrive, based on good maps and data from previous transactions. We could argue that this is just a prolongation of the third revolution, but the way data can be used in the future at least offers the promise of breakthroughs in many fields.

For a retailer, data on what the customer has bought become available as soon as the items are run up at the cash register. These are important for procurement and statistics. However, if the customer pays digitally, it is possible to connect this transaction to an individual. This will also be the case for a cash-paying customer who offers a loyalty card; if not, there will be no way of connecting the sale to the customer.

The value of the transaction data will increase many times over when it can be connected to a customer. Smart software can then make a profile of each customer that can be used for customized marketing, offering customized discounts, add-ons to products, and so on. I discussed some of the related possibilities and limitations earlier.

Environmental Issues

Cash is physical, which requires a system for transporting cash, from the mint to banks, from banks to ATMs, and from merchants back to banks. Most of these transport needs are handled by large armored trucks. In a digital economy, all these transport needs are fulfilled by computer networks. There is no need for trucks that run on gas or diesel.

As we saw Chapter 4, cash has lost its agility in many countries. Thus, most of the cash that is received by merchants has to be transported back to the banks. This can be done by the merchants themselves, carrying the cash from the shop to the bank. Restaurant owners will not find an open bank after their closing time and may drop the money in a night-safe at the bank. However, cash handling is an expense for most banks, which is especially damaging for banks that compete with true Internet banks that do not offer cash functions. In countries that have gone digital, the incumbent banks have closed down cash-handling functions. In practice, their offices will no longer handle cash; there may be an ATM but not a night-safe. When this happens cash must be handled by companies that specialize in cash handling. Since these often do not have local offices, the transport routes will become longer than before.

Zero Interest Bound

During or after a financial crisis both consumers and businesses may try to be careful, saving instead of spending in case the crisis will affect them. To get an economy moving, a central bank may lower interest rates. This makes it more expensive to have money in the bank and the idea is that consumers will spend it instead, in this way activating the economy. It will also be cheaper for businesses to loan money to expand, for example, by investing in new machines and factories.

Lowering interest rates was an important way of reducing the effects of the 2008 financial crisis. The problem is that interest rates have remained very low since then, close to zero in many countries. A few countries have also tried small negative rates for large customers. While the central banks may want to get rates back to "normal" levels, the economy is still struggling and there are fears that a raise will stop the weak positive trend that we are currently seeing in many countries.

Of course, interest rates can technically go below zero. Even today, most customers will lose money on their savings accounts. With interest close to zero and inflation around 2 percent, the buying power of savings will diminish by 2 percent every year. For example, the $100 that you paid into a savings account today will only have the buying power of $98 when you withdraw it next year. Assume that interest is set at -1 percent, implying that you have to pay the bank to keep your savings; then the total yearly cost will be 3 percent.

This will give customers an incentive to withdraw their savings in cash and hide them "under the mattress." There is a strong correlation between low interest rates and the sales of safes.[6] Customers will still lose buying power due to inflation, but will avoid having to pay the bank to keep the funds. Some countries have a wealth tax —that is, a tax that citizens must pay on all wealth, including savings, albeit with a large minimum amount. While the savings in a bank cannot be hidden from the tax authorities, this is easier with cash. In the above example, a 1 percent wealth tax will imply that you only get the value of $96 back after a year if you keep the money in the bank, but $98 if you hide it at home.

In conclusion, most economists accept that zero is a practical lower bound for interest rates. This leaves the central bank with an unloaded gun for the next crisis. In his book, "The Curse of Cash," Kenneth Rogoff analyzed this situation in detail, discussing other ways of encountering the next financial crisis than lowering interest rates. However, there are few good alternatives. Rogoff's solution is to get rid of all high-denomination notes, which would make it very impractical, if not impossible, to convert large sums into cash, making it viable to set interest rates below zero.

Discussion

Societies expend a lot of effort to control crime. At the same time, they uphold a system that encourages crime—namely anonymous cash. The anonymity of cash is of special importance to criminals, but less important to ordinary citizens. This distinction is important. As citizens we will expect to have freedom regarding how to live our lives. At the same time, we are members of a complex modern society. Traveling abroad, we accept that we have to show a passport at borders. Similarly, we accept that police officers are entitled to ask for our driver's license and that we have to go through a security check at the airport. Should we accept that our monetary transactions can be traced? Is this also a necessary part of being a member of a modern society?

My answer is yes. Clearly, I don't want all my data to be published on the net. I assume and demand that banks and other financial institutions, along with the authorities, keep my data secure and secret. However, I will accept that the authorities have a right to calculate the total of my bank accounts, my debt, interest gained or paid, and that they use this information for tax purposes. This is nothing new; the authorities have always had this right, but in a digital economy they may be much more effective.

This will especially be the case for targeting businesses. In a digital economy it will be much more difficult to hide income and to pay employees off the record. Take a restaurant as an example. When guests pay in cash it is easy to keep some of the money out of the ledgers. Tax authorities may get receipts and see if these are registered, but this is a cumbersome operation. In a digital economy, the authorities can perform such an operation automatically. For example, it now becomes possible to check all digital payments performed by customers against the ledger. The transactions are already in electronic form and the tax office can demand to receive the ledger in the same form. Of course, a restaurant and any other business may try to increase the expense part. This will be possible in principle, but when procurements are registered digitally it will be possible to check these against sales. Tax authorities will probably become very good customers for firms that sell data mining software. From then on it will be very difficult to avoid paying VAT or tax on income.

Similarly, it is now possible to check the actual income of all workers. Some transactions may be hidden as repayment of loans or payment for buying personal items. For example, you can pay a carpenter digitally by pretending to buy his car. The risk will be much higher than when using cash. By collecting data from many sources, such as suppliers, and comparing these with well-known averages, algorithms can pick out taxpayers that require manual scrutiny. The problem, both for the carpenter and his customer, is that the electronic payments can then be traced. For example, the alleged payment for the car may be compared to car registrations.

Is this an attack on personal freedom? Perhaps it is just the opposite. In such a system it will be easier to collect taxes from all. Then it will be possible to maintain good social security and health systems, build new roads and airports, improve the educational system, support families with children, and all the other things we expect to have in a modern society. Or, the additional tax income can be used to reduce taxes. There are clear advantages when we can establish systems that are fair to all. It is easier to pay a tax when one knows that everybody else also has to pay their taxes.

This balance between a digital economy and personal freedom is very direct in the fight against terrorists and other criminals. It will be much more difficult for these groups to operate in a digital economy. In particular, it will be easier to follow the money through various systems, perhaps helping determine who is financing terrorists and who is behind organized crime.

While these groups can try to find other non-traditional forms of money, they will, sooner or later, encounter the traditional currency systems. Terrorists need ordinary money to pay expenses. Customers of criminals, people who buy narcotics or pay for prostitution or for illegal workers are participating in the ordinary economy and will have their funds in bank accounts like everybody else. Yes, the illegal transaction may be hidden as something else, but as we have seen this will be very difficult when data from many sources can be analyzed. The problem is then that if a criminal is caught, it will be possible to study all previous transactions and in this way get back to the customers.

It will be necessary in some way to compromise privacy in order to get better security and a more just society. At the same time, the privacy of the victims of crime, such as people who are held up at gun point, are sex-trafficked or become drug addicts will be strengthened. Taking a broader view we could also extend this to include the "privacy" of the children who need a good education, sick people who need health care, or the elderly who need a place in a retirement home. In a cash-free society it will be easier to enforce regulations in the workplace, from salary to job security, and also possible to finance expensive social security systems. All in all, the tradeoff seems to be in favor of citizens.

Notes

[1] Norwegian Hospitality Association (NHO Reiseliv) is the largest employer and trade organization in Norway and has taken this problem seriously. It has paid for several reports that have discussed the problems of undercutting laws and regulation in the hospitality area. NHO Reiseliv sees cash-based operations, from pubs, small restaurants and private lodging, as a threat to its business.

[2] Olsen, K.A. and Staalesen, K. (2013) "Et Kontantfritt Reiseliv," NHO Reiseliv (in Norwegian).

[3] With an exchange rate of 8.6 kroner to a dollar.

[4] Olsen, K.A. and Staalesen, K. (2013) "Et Kontantfritt Reiseliv," NHO Reiseliv (in Norwegian).

[5] There are some discrepancies here. Some consider the second Industrial Revolution to have started with the invention of the steam engine (around 1870), while others wait for Henry Ford and true mass production.

[6] http://fortune.com/2016/02/23/japans-negative-interest-rate-driving-up-safe-sales/
http://nordic.businessinsider.com/negative-interest-rates-causing-safe-sales-spike-2016-6?r=US&IR=T

Chapter 13
Disadvantages of a Digital Payment System

While most people find clear advantages to replacing cash with digital payments, some may see disadvantages. The main difference between these two systems is that one guarantees anonymity while the other may be traceable. Untraceable digital transactions may be achieved with cash cards or virtual money. However, as seen in the previous chapter, we would expect to replace cash with a system that is traceable. Apart from anonymity, a cash-free society requires that everyone has a bank account and access to the different payment systems. Without cash, the digital systems must handle the needs of everyone.

I shall try to address all of these issues, including the disadvantages and how they can be reduced.

Loss of Anonymity

Most ordinary citizens will have little to fear from the traceability of digital transactions. The fact that the transactions are traceable in principle does not imply that someone will take advantage of this possibility. We can assume that there are security mechanisms in place to hide all data from unauthorized persons. There is, of course, always the risk that your spouse may notice how much wine you buy, the real expense of your new fishing rod, or the hotel bill that you want to keep hidden. In practice, this can be handled within the digital economy, for example by having an extra credit card account.

As we shall see in Chapter 14, Norwegian citizens do not seem to be concerned about missing some degree of anonymity. In a survey of a thousand people performed three times between 2013 and 2017, only 4 percent of those who use cash indicated that they did so for reasons of anonymity (the figure was the same in each survey). In many ways, this is a reasonable answer. They are on Facebook and Instagram, carry a smartphone where their location can be detected at any time, use email, and shop online; in other words, they use systems where maintaining strict anonymity is not easy.

Some people are afraid that the government may use digital payments to control the population. This is especially the case in countries where inhabitants do not trust their government or where there has been a history of previous dictatorial governments. In such cases, IT technology, surveillance equipment, smart algorithms to oversee Internet traffic, and digital payment systems may all be weapons that can be used against the population. History tells us that occupiers of a conquered country often try to control the economy by issuing their own currency. This was done by the Nazis in occupied Europe during the Second World War. Today we can expect that an occupier would establish a digital economy to replace cash. This can be done regardless of whether there was a digital economy at the beginning or not, so one cannot expect to protect the anonymity of citizens during an occupation. If the occupier wants to control the economy, transportation, movement of people, and so on, there are many opportunities to do this.

While laws against occupation or creating a dictatorship may not help, there should be safeguards in place. In democratic countries we expect an independent judiciary to maintain the right to privacy. A free press is important, especially one that has resources to investigate matters (which may become a problem in the future as newspapers and other news agencies have problems maintaining their business models). Also, it is important that there are mechanisms in place for erasing data on previous transactions. This should ideally be performed continuously; for example, erasing all transactions that are more than a year old. It would also protect the citizens if all these data were erased when a country was occupied. If not, the occupier could use data-mining techniques to pick out people that were considered dangerous. However, in this respect many data sources other than just economic data could be misused if captured by the occupying force.

People without Bank Accounts

In the United States, 7 percent of households do not have a bank account and 25 percent do not have a credit card.[1] The numbers from the OECD countries are similar. In the Scandinavian countries the situation is very different; here, virtually all adults have a bank account. On a global basis, it has been estimated that 38 percent of people did not have a bank account in 2014, although this number decreases every year— it was 49 percent in 2011.[2]

Some countries demand a minimum income and also impose charges on an account, but this is not the case in Scandinavia, where banks cannot deny a citizen a bank account. While people with a history of credit abuse may have a problem establishing an account in some banks, others may offer an account with a debit card to such groups.

Clearly, for an economy to go digital, everyone needs to be on board. This can be achieved by forcing the banks to offer accounts and cards, by subsidizing banks to do this for low- or no-income customers or by allowing members of the latter group to open an account in the central bank. Citizens who do not participate in the banking system today will probably be among the ones that will benefit most in a digital economy. Instead of having to pay high fees for cashing checks, they will now be a part of a zero-fee or close-to-zero fee system. The advantage for authorities is that welfare can be paid directly into accounts.

There is always the option of using cash cards. Some countries offer welfare on these cards. The disadvantage is that anonymous cash cards are a way of establishing a new anonymous payment system. It also seems more reasonable to offer a general debit card to everybody and let them participate in the common economy instead of using special solutions for some groups.

Usability

Everybody can handle cash. With digital payment systems, people have to learn new procedures for paying. Instead of just looking in your wallet to see how much money you have available, you need to access an account.

However, when a digital system is set up, when the customer has received the debit or credit card, or when the appropriate app is installed on a smartphone, the actual payment process can be made to be very uncomplicated. For small amounts, often below US$40, the tap-to-pay system is the simplest option. A user just holds the card or the smartphone next to the reader and the payment is authorized. For larger amounts, a PIN or a fingerprint may be required.

People who cannot remember their PIN may have a problem if the amount exceeds the tap-to-pay limit. But turning to cash will not help. In many countries, banks are leaving cash handling to ATMs. That is, one needs to provide the PIN just to get cash. There are ways of overcoming this problem and I shall return to this in Chapter 15, where I provide examples of nice-to-have functions that may exist in the future.

Digital payments work best if the user has access to account information on the Internet. This also simplifies the task of paying invoices. While more and more people are learning to master these services, there are some—often senior citizens—who do not have the necessary skills to use an Internet bank and may also lack the equipment needed to do this. While banks can send account information as letters in the post, perhaps requiring an additional fee for this service, these non-computer-literate groups will have a disadvantage in the digital economy.

In practice, many people find ways to overcome this problem, for example by asking someone in their family for help. In some ways Internet banking makes this even easier than before, as the helper can be located anywhere. If this is not a viable option there will always be banks that can provide a personal service, but as we see today, the services that were free some years ago may now incur a fee.

Exceptions

Cash is simple for small payments. A kid can go down to the corner store with some coins to buy a chocolate or an ice cream. You can pay back a small loan to a friend in cash. We can use cash to pay for strawberries or apples on the side of the road. The farmer does not need any form of equipment to handle these transactions. We see that cash transactions with small amounts are simple, as long as the cash is available. Drivers along the road who do not have cash available will not be able to buy anything from the farmer.

A modern digital payment system will handle all these instances. In practice we can use the standard debit or credit cards. First, there should be no minimum amount for using a digital transaction. When the system is set up, with terminals and broadband, the cost of yet another transaction may be very low. As discussed in Chapter 7, there should ideally be no transaction fees for customers and only minimum fees for merchants. For the customer, it is most convenient when the same payment system, whether it is a card or a smartphone, can be used for every transaction independent of the amount.

The farmer can choose to install a terminal. Another option is to get a card reader for his smartphone. These are inexpensive, and sometimes also offered for free, but there will be a charge of 1.5–3 percent on each transaction, depending on the total volume. The third option for the farmer is to rely on the customer's smartphone—that is, to treat the transaction as a person-to-person payment.

Person-to-person payments can be performed easily in a digital payment system, often better than with the use of cash. With a bank app on a smartphone, the amount can be paid as an account-to-account transaction. This requires that the account number is known. In some systems, one can ask for an immediate execution, but in most cases these transactions will be executed only at given times. Then the customer has the possibility of deleting the transaction before it is executed. However, many Internet banks now offer an option for immediate transactions.

New mobile payment systems, developed specially for person-to-person transactions, use the telephone number of the receiver. These transactions are always performed immediately and the receiver receives an instant notification that the funds have been received. This makes these systems convenient for buying and selling. In the end, the amounts are taken out and inserted into bank or credit card accounts. The advantage of using digital payments here is that everything is recorded. The question "did I remember to pay?" can easily be answered by opening the app.

Children going to the store can have their own card or use their smartphone. Experiences from countries that allow small children to have debit cards are very good. Now banks are promoting special cards for the very young. The parents sign up for these cards and also transfer amounts to the accompanying accounts. The cards are usually free to buy and involve no charges.

Tourists

A country in the digital economy may get visitors from countries where cash is still king. However, most of these visitors will have a credit card. This is the experience from Norway, where most banks have eliminated cash handling and currency exchange. A few restaurants or hotels allow international customers to pay in dollars or euros, but this becomes more cumbersome as the merchants do not have an easy way of getting the cash into their bank accounts. In practice, a credit card is necessary if you want to visit Norway. For example, if you travel by road, some automatic toll stations will require a credit card.

For those who do not have a card, the solution must be to establish a temporary debit card account (a "tourist account") that allows the customer to charge the account with cash. This could be done at airports, replacing the cash-to-cash interchange that is available today. Still, the tourist will probably need an ordinary credit card to rent a car and to book into most hotels.

Paying the Babysitter

Today we can pay the babysitter or the person who cuts the lawn in cash. We can just as well use a digital person-to-person transfer. But since this is traceable, we may ask if the payments should be registered as a salary; should the service provider pay tax and social security fees?

First, the possibility of tracing a transaction does not imply that anybody *will* do this. We should expect that small amounts now, as before, will go under the radar. However, it would be better that these "grey zone" situations are made into law and regulated. This can easily be done by defining that income below a certain threshold is tax-free and that "salary" in the same manner can be paid informally. Then it does not matter if the payment is digital and in principle traceable. However, if the babysitter does this as a living, he or she may find that tax authorities become interested.

This takes us back to the issue of formalization that I discussed in the beginning of this book. When moving from an informal system, such as using cash to pay for small services, to a more formal one, using traceable payments, it will be necessary to adjust the laws and regulations. In many cases one can use the method described above, of redefining the border between black and white, thus removing the formerly accepted grey zone.

Grey Zone Payments

Cash can be used for legal monetary activities and for criminal activities. Between these extremes there is a grey zone. For example, buying mild narcotics such as marijuana is illegal in some societies, but if the police do not try to arrest small-time sellers and users, it can still be accepted in practice. Cash then becomes a convenient form of payment. With digital transactions, authorities will have the means to follow up on both buyers and sellers, but there is no requirement for them to do this. If they don't, the situation will be the same as when cash was used: no one interferes. However, it makes a difference when every transaction is recorded. The political situation may change. A new police chief may want to go back through the records to find the offenders.

Technically, citizens can be protected by deleting transaction records after a certain time, but it may be better to change the law so that it is more in accord with reality. We see that several states in the United States, such as Colorado, have legalized the sale of marijuana. However, this adjustment of laws with regard to public opinion is not synchronized with the rest of the country. Selling and buying marijuana is still a felony at the federal level. Thus, banks and credit card companies that operate all over the US are afraid of having anything to do with the marijuana shops. They are frightened of being prosecuted by federal law enforcement, so they rely on cash. Logically, this should imply that the central bank, the issuer of the dollars that are used, should be prosecuted. The difference is that it is easy to see that a debit or credit card transaction originates in a marijuana shop, while cash can be used for so many activities that there is no way of controlling its use.

Again, we see a clear example of what happens when we formalize a system. Using cash to buy milder drugs may be tolerated since cash can also be used to buy milk and bread. Even when the cash use is way beyond the situation in other states (an indication that it is used for buying marijuana), the anonymity of cash allows the federal law enforcers to look the other way. This offers the interesting situation where it is the authorities that welcome the anonymity of cash.

While federal and state laws may eventually be synchronized, there could be several problems with "formalizing" the payment systems. I discussed these issues in Chapter 2. If the ideal solution cannot be found right away, we have to live with approximations. If the marijuana-buying citizens of Colorado could not use cash, there would clearly be an incentive to establish an in-state bank that provided the necessary on-line service, hoping that this bank was not being prosecuted by federal law.

To limit the consequences of the transition from a cash-based to a digital economy, it would be smart to let the change run over many years. This would allow time both to detect and correct problems of formalization, which occur when replacing a very flexible system (here cash) with a somewhat more transparent system (a digital economy). This flexibility can be implemented by starting to remove the high-denomination bills.

Ideally, we may end up with a society of only necessary laws and where all of these are enforced.

Loss of Seigniorage

Seigniorage, discussed in Chapter 5, is the money that the issuer of the currency will receive. The value engraved on a coin may be greater than the cost of producing the coin; the difference is seigniorage. For banknotes, this difference can be significant. In fact, each banknote is a document of debt, indicating that the central bank is willing to pay the denomination of the note to the holder. In practice, these "debt certificates" are trusted in a population so that they can be used for payments. The advantage for the central banks is that these debts are interest-free; also, many of the notes issued will never be redeemed. This is especially the case for currencies such as dollars or euros that are used all over the world.

In a digital economy, seigniorage will disappear, at least in the form that we have discussed here. This will represent a serious loss of income for central banks and may explain why central banks try to uphold cash. We shall discuss these issues in greater detail in the next chapter, using the Norwegian central bank as an example. However, another form of seigniorage is "opportunity cost seigniorage." This is the idea that the central bank can issue debt as an alternative to printing money. This will still be a possibility in a digital economy.

In the previous chapter we saw that the amount of money that the central bank allows into the economy must be balanced. If it is too little the economy may be weakened and make a recession more severe; if too much money is allowed, inflation may increase. Some of the money, especially the part that comes as cash, may not be put into the general economy as citizens may store this "under the mattress." This is a temporary storage, but if many keep their values stored away there may be room to insert more money into the economy. However, if inflation goes up there may be an incentive to use more of these stored values before they depreciate. Balancing the amount of money that goes into an economy may be compared to dragging a string over a table; it may accelerate and deaccelerate even if we try to drag it with an even force.

Some currencies, such as the US dollar, have been a means of conserving value in many developing countries. Today the euro is also used in this manner. These currencies have the advantage that they are used for payments along with national currencies. There is often a rebate when paying with dollars—that is, a better exchange rate than the official rate. This is an advantage for the US economy. In fact, the US Treasury can "sell" dollars that are very cheap to produce for their face value, receiving goods and services back. Most of these dollars will be kept and used in the countries that receive them. A few may come back to the US, for example in the hands of tourists, but most will never be returned; there is little chance that someone will ask for all the "debt" to be repaid. In this respect, using dollars either for paying or for storing value in an underdeveloped country is a way of subsidizing the United States. The EU is in a similar position.

The amounts involved are staggering. In a paper from 1997,[3] Edgar L. Feige calculated that, since 1964, "the cumulative seigniorage earnings accruing to the U.S. by virtue of the currency held by foreigners amounted to $167–$185 billion and over the past two decades seigniorage revenues from foreigners have averaged $6–$7 billion dollars per year."

In a digital economy, monetary seigniorage will be a lost income for the central bank. In addition, one may have to buy back the cash that is no longer needed. However, the transition to a digital economy will run smoothly independent of the central bank. We have already seen this in the Nordic countries, where cash is now only used for a small percentage of transactions. This development will continue until cash is the exception in everyday transactions. The process will be even smoother if the central bank takes control and withdraws the high-denomination bills from the market. This can be done step by step, for example, in the United States starting with the $100 bill. In smaller countries, where only a minor part of the currency is held abroad, the withdrawal of these bills will be even easier.

Revoking the high-denomination bills will also create opportunities to get rid of much of the money that is in the hands of criminals. All large sums of cash that are turned in could be required to be accompanied by a statement telling where the cash originated and where it has been stored. This will offer a possibility to tax values that have been hidden from the authorities. It is therefore likely that some of the bills will never be turned in. While one may accept paying a heavy tax on hidden funds, it may be more difficult to provide a reasonable explanation of funds that originated from illegal activities.

By removing one denomination bill at a time, the transition from cash to digital should be a smooth operation. The central bank will have to issue debt certificates, short- or long-term treasury notes, to get the funds necessary to buy back the cash. Replacing zero-interest loans such as cash with interest-bearing certificates will reduce the income of the central bank. In addition, there will be no seigniorage from cash in the future. The central bank will probably then have to acquire funding from other sources. In the meantime, this loss of income may be an incentive for the central banks to maintain cash as long as possible.

In a true digital economy, it will not be possible to keep money "under the mattress." Instead, citizens will put their money in a bank account or use it for investments. That is, the money will be in the economy where it can play an active role. For example, the bank can loan the money to a company that is expanding or to a family that wants to build a new house. In the long run, moving money from "mattresses" into the economy will create jobs and the government will be able to tax the money.

Digital Security and Cash as Backup

Cash is often seen as a possible backup when there are power blackouts, when data networks are down, or when the computers that handle the transactions fail. This may be true when a large part of the population uses cash, but in most modern countries cash is now on the way out. If we look at monetary value, only an insignificant part is in cash, which means it is more doubtful that cash can be a real backup.

Some countries have strained power grids. Due to a weak economy, many power companies use insufficient funds for maintenance and for building redundancy into the grid. This creates a risk of prolonged blackouts, as both the United States and Canada have experienced in later years. Battery backup may keep some systems running for a few hours, perhaps during shorter backups, but for prolonged blackouts there will be no chance of maintaining service, except for critical institutions that may have generators powered by diesel engines.[4]

Many years ago, when blackouts were more common, merchants were able to cope with blackouts; lighting gas lamps would provide enough light, with a handle, an electric cash register could operate in manual mode.

Today this is very different. No modern computer network, terminal, or digital payment system will work during a power blackout. Even with cash there would be severe problems. A customer will encounter the problem when the electric door of the supermarket will not open, and even then the door would enter into a shop with no lighting. Groceries need power for fridges and freezers. They may be able to sell you other goods, but the cash register would not work. In many grocery stores, prices will not be available, as these are stored in databases and not on the products themselves. A gas station may still be able to take cash, but without power the pumps will not work.

So let us assume that we have power but that the networks or the central systems that handle payment transactions are down. At the grocery store, one may still find that prices are no longer available as the central databases are not accessible. However, with local backup there is a chance. Those that have cash will be able to pay the cashier until the cash runs out, but others will have a problem. If computer networks are not running, the ATMs will not work. Even in countries where banks still have cash services they cannot let you withdraw funds without access to central registers.

Thus, cash will not work as a backup unless authorities take steps to ensure that merchants and banks can operate independent of central databases, that there are local backup of prices and other data. It is also important to ensure that there is cash available and that this can be distributed to citizens even when data networks are not up and running. Citizens could be required to keep a cash amount as a backup, but this will not be practical. Many will not have the funds available and most will have problems finding a secure place to store the money. As we have seen, cash is not a viable backup in a modern society.

Many terminals may operate in offline mode—that is, transactions are stored locally until the networks are working again. While the terminal can perform a simple verification of the card, there will be no possibility to check whether there are funds available, for example in a debit card account. While this is a good option when the payment systems have problems for a limited time, it becomes more difficult for prolonged down-times.

The only viable solution is to build secure systems using redundancy—that is, systems that run even if separate components fail. A good example is modern planes; these are technically complex and the safety of modern air travel is partly due to redundancy. A modern plane has two or more engines, two pilots, two independent control systems, and so on. If one engine breaks down, the plane can go on using the other. The probability that the plane will have problems with one engine is very small; the likelihood that two engines will fail at the same time is extremely small.[5]

A digital economy can be secured using the same technique. Redundancy can be achieved by investing in more equipment so that there are at least two independent systems. Often this redundancy can be achieved by reaching agreement between competitors. For example, if the mobile network for phone company A is down, customers should be able to call on companies B or C. To handle the increased traffic on the other networks, company A's customers could face restrictions on the other networks, for example by only accepting short telephone conversations or limited data transfers.

Power backup can be provided with independent supply systems. This may require more capacity, additional transmission lines, new battery backup systems, and so on, and may come at a high expense, but may be cheaper than taking the cost of prolonged blackouts. Critical parts of the society, such as hospitals, mobile networking towers, and other communication networks, should have independent power backup, from batteries to diesel generators. New battery technology offering high capacity may provide good backup solutions for temporary blackouts.

One cannot expect that all of the various operators, such as phone companies, network providers, or banks, will invest in this additional security in a competitive environment. Nor will it be a good idea for authorities to specify the technical solutions. What the authorities can do is to fine the providers for every minute of downtime. If the fines are heavy enough, the providers will find good ways to avoid downtime.

On the other hand, citizens should ensure that they have some form of power backup, such as battery packs for charging mobile phones or the ability to charge a phone in the car. Electric cars, especially hybrid ones, could also be a source of limited power backup for appliances such as a fridge or a freezer.

Discussion

When the discussion about digital versus cash payments first came up many years ago, it included a long list of digital disadvantages and cash advantages. The "what about" questions included senior citizens, small-time merchants, children, security when paying in a bar, slow digital systems, and many more.

As we have seen today, all of these cases have been or are in the process of being addressed. Senior citizens have to use a card to get cash and may then just as well use the card in the stores. There are inexpensive payment solutions for small-time merchants, from card readers that can be connected to smartphones to systems for real-time payments using just a smartphone. Banks provide debit cards for children as young as eight, and younger children may use cash cards. We have seen that it is easier to pay digitally in pubs, especially with tap-to-pay solutions. Today, when terminals have a broadband connection, digital payments are faster to perform than cash payments. Of course, these services are not available in all countries and cash will not disappear before the digital payments systems can handle all types of payments. But many of the advantages of going digital will become apparent long before the last coin is minted. As we have seen, simply removing high-denomination bills will diminish the use of cash in the underground economy.

It seems that all except one of the disadvantages has been eradicated: digital transactions are not anonymous. If a society is going to reap the advantages of digital payment systems with regard to fighting terrorism, making life harder for criminals, retrieving taxes, and regulating the workplace, full anonymity cannot be offered. This does not imply that payment data will be available to all, just that the authorities will have access. The cost of this loss of transaction anonymity will be very low for law-abiding people and very high for criminals. These arguments are only valid in a democratic society where the government does not misuse its power. If this is not the case, it will not be realistic to discuss the merits of anonymous cash, as any undemocratic government can choose to abolish cash in order to get better control.

Notes

[1] Federal Deposit Insurance Corporation, data from the 2015 FDIC National Survey of Unbanked and Underbanked Households. (FDIC)https://www.fdic.gov/news/news/speeches/spsep0816.html.

[2] *The Economist*, September 9, 2017.

[3] Feige, E.L. (1997) "Revised Estimates of the Underground Economy: Implications of U.S. Currency Held Abroad," in O. Lippert and M. Walker (ed.) *The Underground Economy: Global Evidence of Its Size and Impact* (1997), pp. 151–208. https://ideas.repec.org/p/pra/mprapa/13805.html

[4] Newer methods of battery storage, such as those produced by Tesla, may offer an alternative to diesel generators in the future.

[5] This happened in 2009 with US Airways Flight 1549, which had to land in the Hudson River when both engines became useless after a bird strike. Still the captain managed to land the aircraft in the Hudson without any loss of life. This was possi-

ble since the captain could start a third engine, an Auxiliary Power Unit (APU). The APU gave no thrust but provided power to the instruments and to the hydraulics so that it was possible to steer the plane. We see that redundancy makes air traffic safe.

Chapter 14
Case: Norway

Norway provides a good case of a country that has moved a long way in the direction of a cash-free society. We could just as well have selected Sweden, Denmark, or Iceland, all of which are stable and homogeneous countries with a low crime rate. Tax rates are high and so are the benefits. Citizens of these countries enjoy free health care and free education. There is a pension for all and elderly people who are in need may get a place at a retirement home. Norway and the other Nordic countries have excellent data networks. Most inhabitants and businesses can receive broadband fiber connections. The mobile networks cover all cities and nearly all roads and communities. In population centers, one may expect 5G or 4G service, while other communities still have 3G, and some have only 2G service.

Norway has five million inhabitants. Banks are required to open a bank account for anyone over eighteen who asks for one unless there is a good reason to deny such a request—for example, if the customer has a long history of credit infringements. Even then, it may be possible to find a bank that will provide such a person with an account and a debit card. In most cases this is without risk for the bank, as there will be an online lookup of the account balance with each card transaction.

In general checks were phased out many years ago. The few that still write checks have to pay heavy fees. The previous methods of phone banking (from a fixed line) and paying by sending a transaction form to the bank by post are now rarely used services. This leaves Norwegians with two active forms of payments: a digital one and a physical one, which is cash.

All businesses in Norway have to accept cash (in theory, also including the online stores), but there is pressure to change these regulations. Many organizations, along with the tax authorities and central police organizations, want to have a system in which all businesses must accept digital payments and can choose whether to accept cash or not. In practice, the requirement to take cash is hollowed out by the new digital applications, net shopping, apps for buying tickets, and more.

However, the country's central bank (Norges Bank) and Department of Finance want to maintain current regulations. Some political parties have suggested phasing out cash, but are meeting with severe opposition from citizens that want to continue using cash, so it does not seem that this situation will get a political solution. I shall return to this discussion in the summary, also explaining why the central bank wants to maintain cash. Summing up, everything is in place for cash-free payments in a country such as Norway. The question is then—what happens?

Compared to Other Countries

Here I shall look at digital payments in a set of developed countries and see where Norway fits in. The countries that I have selected are all candidates for being cash-free. In most of these, the technological infrastructure is in place, but there are still great discrepancies in the use of digital payments.

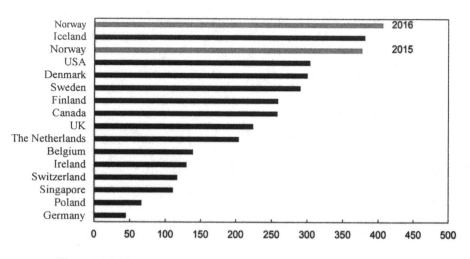

Figure 14.1. Number of card payments per citizen in selected countries
(data from Norges Bank, 2017[1])

Figure 14.1 shows the number of payments made by credit or debit cards in selected countries. Norway and Iceland top the list, with more than 400 card payments
each year for each citizen. Interestingly, countries such as Germany and Switzerland
are quite far down the list.

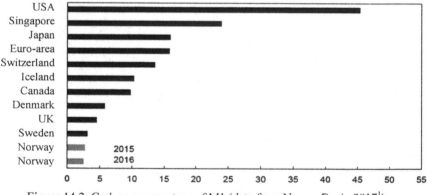

Figure 14.2. Cash as a percentage of M1 (data from Norges Bank, 2017[1])

M1 is the total money available for purchases. It includes all cash and also readily available money in bank accounts (such as checking accounts). Figure 14.2 shows
cash as a percentage of M1. Note that the calculation of M1 may vary among the
countries. As the figure shows, Norway was at the bottom of the list in 2015 and 2016,
with cash representing less than 3 percent of M1. At the other end of the spectrum is
the United States, where cash is close to 50 percent of M1. However, since US dollars
are used all over the world, they will come out with a larger percentage compared to

some other countries. Estimates show that cash was used in 32 percent of all retail payments in the United States in 2015, while electronic payments represented 59 percent.[2]

On a global basis, a market research organization called Euromonitor International reports that consumer card payments surpassed cash payments for the first time in 2016. The compound annual growth rate (CAGR)[3] of card payments and mobile commerce is expected to be 23 percent between 2016 and 2023, whereas global cash payments will have a CAGR of 1.3 percent. A senior analyst at Euromonitor said, "This stagnant growth of cash payments signals a shift from an increase of cash supply to a decrease and is a major victory for card and electronic payments, and the shift of consumer shopping channels toward online is directly benefiting mobile-commerce, which saw a 53 percent rise from 2015 to 2016."[4]

Thus, it seems that other countries are following the leaders toward a cash-free society. It is interesting to present the leaders, as these may give an indication of the future for all countries. As a leader in ecommerce, Norway makes a good case study.

Credit and Debit Cards

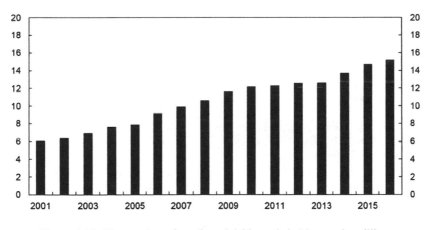

Figure 14.3. The number of credit and debit cards in Norway in millions
(data from Norges Bank 2017[1])

Figure 14.3 shows the number of payment cards from 2001 to 2016. There is now an average of three cards per inhabitant of Norway.

Individuals may have to pay a yearly fee for a debit card, while most credit cards have no annual fee. There are usually no charges for customers to use the cards, but merchants have to pay a fee of 1.5–3 percent of the amount when a customer pays with a credit card, or more for some cards. The transaction fee on a debit card is just a few cents.

Number of Terminals for Payments

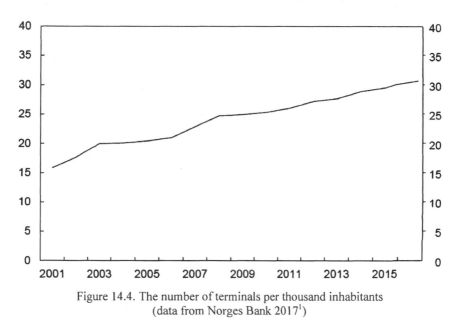

Figure 14.4. The number of terminals per thousand inhabitants
(data from Norges Bank 2017[1])

As Figure 14.4 shows, the number of terminals for payment is increasing stead-
ily. In 2016 there were more than thirty terminals for every thousand inhabitants.

But this is not the full story. By the beginning of 2016, 81 percent of Norwegians had a smartphone, up from 46 percent in 2011.[5] For those under fifty years of
age, the figure is 95 percent. A smartphone can be used for payments in many ways.
There are now apps available for buying tickets for buses, trains, concerts, theater,
and cinema. Mobile phones are used for paying for goods sold on the net. In addition,
person-to-person payment systems are now emerging, and phones can be used for
traditional Internet banking and for paying in a store. Thus, in practice there will be
more than one "terminal" for each inhabitant.

Card Payments

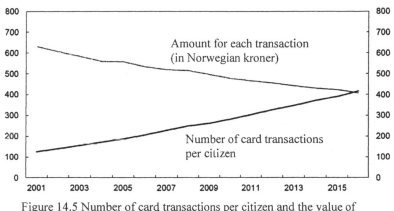

Figure 14.5 Number of card transactions per citizen and the value of each transaction in NOK (data from Norges Bank 2017[1])

Norwegians use their debit or credit cards a lot. Figure 14.5 shows the number of card transactions per citizen from 2001 to 2016. We also see that the value of each transaction decreases, from more than NOK 600 (approx US $75) in 2001 to NOK 400 (approx $50) in 2016 which implies that the cards are used also for lesser amounts. When tap-to-pay systems are becoming more common we should see a continued reduction in the amount per transaction.

Paying Invoices

The three common forms of paying invoices are using Internet banking, setting up automatic payments, and using a mobile phone. An invoice can be paid by filling out a payment form offered by the Internet bank. This is very similar to what people used to do when filling out a check or the transfer forms that were sent by mail to the bank. However, the user interface of the Internet bank can help the customer find the correct recipient.

With electronic invoices, one avoids the data entry part. This implies that the invoice is also sent to the customer's bank account. A notice of a nearly due payment can be sent by mail or email. These invoices can be paid by the click of a button, since all the necessary data is included in the electronic invoice.

A fully automated solution is also available. Many customers set up such an agreement with utilities and other companies that they trust. Invoices are paid in full on the due date; no action is needed by the customer.

Mobile phones are now used mostly for person-to-person payments, but many merchants now offer tap-to-pay solutions.

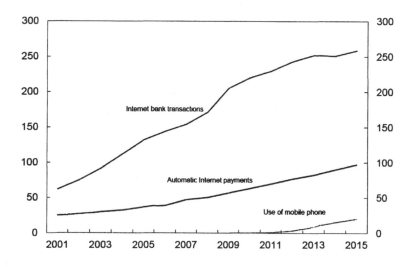

Figure 14.6. Transactions by personal customers. Millions of transactions
(data from Norges Bank 2015[6])

    The volumes of the different payments in millions of transactions (e-Invoices
are included in the Internet transactions) are presented in Figure 14.6.

Branch Offices Close Down

Incumbent banks once had branch offices all over the country, but customers are now
servicing themselves online. There is little need to go into a bank office. The CEO of
DNB, Rune Bjerke, told reporters in January 2017 that DNB, Norway's largest bank,
had ten thousand employees at the time, and that it would be a miracle if they had
more than five thousand in five years. He calls DNB a technology business.

    This development is also forced by the increase of true Internet banks that do
not have any branch offices at all; these have taken a large share of consumer banking
in Norway. The incumbent banks believe that they may also have to compete with
companies such as Amazon, Apple, Google, and Facebook. While the banks could
have chosen a strategy of retaining offices and offer a "personal experience" that the
true Internet providers could not offer, most have chosen instead to be fully digital—
that is, to meet the competition without large expenses for maintaining branch offices.

    In this choice of different strategies, one must follow the customers. Trond
Bentestuen, a director at DNB, says that customers no longer need to go into a bank
office to contact the bank. Instead they use their mobile phones and PC. In fact, the
bank has more contact with customers today than previously, but nearly all the traffic
is digital.

Year	Number of head and branch offices
2008	1330
2009	1184
2010	1157
2011	1158
2012	1127
2013	1061
2014	1042
2015	991
2016	953

Table 14.1 Number of bank offices in Norway (data from Finans Norge)

We see the development from Table 14.1 which presents the number of bank offices in Norway at the start of each year, both head offices and branches, from 2008 to 2016. The numbers show that that nearly one out of ten bank offices is closed down every year. DNB reduced its number of branch offices from 116 to 57 in just the first half of 2016. This is in addition to seventy further branches that have been closed in recent years.[7]

Bank Offices without Cash

Cash handling is expensive for the banks. In order to meet the challenges for the true Internet banks and other digital providers, they try to limit the cost of this traditional service with different policies, such as the following:

1. Reducing the number of branch offices
2. Removing cash handling from the remaining branches
3. Directing customers to ATMs
4. Reducing the number of ATMs
5. Imposing a fee for ATM operations
6. Outsourcing ATM operations

As the banks close their offices and remove cash operations, customers are being directed to the ATMs. However, the number of ATMs is declining.

From January 2017 DNB applied a fee of 10 kroner ($1.20) for cash withdrawals at an ATM. Other banks are introducing fees as well. The fee is not high, but it has broken a traditional understanding between banks and customers that cash handling should be free. Interestingly, these fees have been accepted without much discussion. The reason is that most of the customers seldom use ATMs, if at all.

Some banks in Norway have moved away from cash handling altogether by letting other companies take over their ATMs. The new operators are companies that

specialize in handling cash, transporting cash from businesses to banks, offering
change, and filling up ATMs.

Number of ATMs

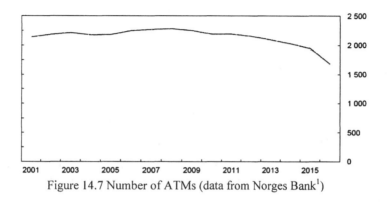

Figure 14.7 Number of ATMs (data from Norges Bank[1])

Figure 14.7 shows that the number of ATMs is declining, at an increased rate from
2015 to 2016.

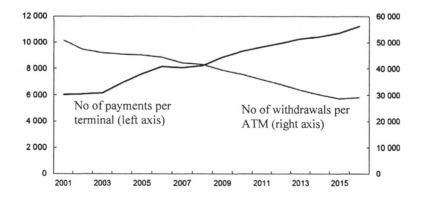

Figure 14.8. Use of terminals and ATMs (data from Norges Bank[1])

This is also the case if we look at usage of each ATM. As seen from Figure 14.8
the usage of each payment terminal is increasing, while the number of withdrawals
from each ATM is declining.

Cash as a Percentage of M1

M1 is a measurement of the most liquid portions of the money supply in a country. It includes the type of money used for payments, cash, and funds in bank accounts that can easily be used for payments or for being transferred into cash.

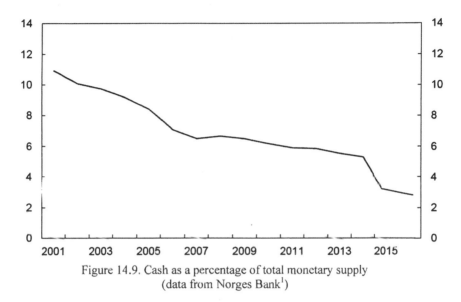

Figure 14.9. Cash as a percentage of total monetary supply
(data from Norges Bank[1])

As seen from Figure 14.9, cash is gradually being taken over by account money and cash now represents only 2.6 percent of the total monetary supply.

What Do the Citizens Say?

The Norwegian Hospitality Association (NHO Reiseliv) is a part of the largest employers' and trade organization in Norway. Several of their members have to compete with hotels, restaurants, and pubs that cheat on taxes and that do not pay their employees according to regulations. Therefore, NHO Reiseliv is advocating a transfer from cash, which opens for underground activities, toward digital payments.

With funding from this organization, I performed three surveys, in 2013, 2015 and 2017. In each of the surveys, one thousand people were asked about their payment preferences.

The first survey was directed toward payments in the hospitality business, while the 2015 and 2017 surveys are more general. All polls are performed by NORSTAT, a survey company.[8] The result of the surveys is presented as a part of three reports, all of which are in Norwegian.[9] The main results are presented here.

In the following tables, the 2017 numbers are presented on top, in normal font, while the 2015 and 2013 numbers are presented in italics.

Chapter 14

Use of Digital and Cash Payments

			Age				Sex	
	TOTAL	18-29	30-39	40-49	50 +	Male	Female
I only use digital payments	37 %	38 %	47 %	43 %	32 %	40 %	35 %
2015:	30	27	38	36	25	39	21
I mostly use digital payments	46 %	53 %	45 %	46 %	42 %	42 %	50 %
2015:	54	61	57	52	50	44	64
I use digital and cash equally	12 %	6 %	6 %	6 %	20 %	12 %	13 %
2015:	12	9	3	9	19	12	13
I mostly use cash	4 %	3 %	3 %	4 %	4 %	5 %	2 %
2015:	3	2	1	2	4	4	2
I only use cash	1 %	1 %	0 %	1 %	2 %	2 %	1 %
2015:	1	0	1	0	2	2	1
TOTAL	100 %	100 %	100 %	100 %	100 %	100 %	100 %

Table 14.2 Which of these are most correct for how you pay in
shops, hotels, restaurants and other places?

Table 14.2 shows that 37 percent of respondents are only using digital payments, such as debit cards, credit cards, and smartphones, an increase from 30 percent in 2015. The increase is highest for females, which have increased from 21 to 35 percent in these two years. Nearly half of people in the 30–39 age group use only digital payments. In practice, there are no problems being a part of this group as nearly all merchants take cards and mobile payments, there are no lower limits and no fees.

In the 30–39 age group, 92 percent are "digital most" or "digital only." For all age groups this number is 82 percent for men and 85 percent for females. Most of the increase in the "only" group has come from the "most" group. The changes in the other groups from 2015 to 2017 are minor.

The cash-"only" group is at 1 percent. Even merging "only cash" and "mostly cash" only adds up to 5 percent.

I should note that these data were gathered before a set of new person-to-person money transfer services have been introduced in Norway. In addition, tap-to-pay cards have just been introduced. Therefore, one should expect the numbers for "digital only" to increase in the coming years with a similar decrease in the cash groups.

The card can be misused		6 %
	2015:	*3*
It is simple		32 %
	2015:	*28*
I don't use cards		1 %
		4
I have more control of what I use		28 %
	2015:	*23*
It is anonymous		4 %
	2015:	*4*
I have income in cash		5 %
	2015:	*4*
Other		45 %
	2015:	*60*
Don't know		4 %
	2015:	*4*
TOTAL		124 %

Table 14.3 What is the reason that you pay with cash?

The survey then asked those who use cash why they chose this form of payment (Table 14.3). While the unspecific "other" was the largest response, 28 percent indicated that they feel that they have more control with cash. I shall return to this in Chapter 15. Note that this group is quite small.

Only 4 percent of respondents indicated they use cash because it is anonymous; this number is consistent with results from 2015 and 2013. In the defense of cash, anonymity is often presented as the main argument, but these surveys do not back up that claim.

Paying in a Pub or Bar

		Age group				Sex	
	TOTAL	18-29	30-39	40-49	50 +	Male	Female
Digital (card or phone)	80 %	86 %	86 %	83 %	74 %	76 %	84 %
2015:	71	76	79	75	64	71	70
2013:	53	75	59	43	44	53	53
Cash	20 %	15 %	14 %	17 %	26 %	24 %	16 %
2015:	29	24	21	25	36	29	30
2013:	47	25	41	57	56	47	47
TOTAL	100 %	100 %	100 %	100 %	100 %	100 %	100 %

Table 14.4 Paying in a pub or bar

In the 2013 survey it became clear that many customers were not comfortable paying digitally in a pub or bar. While 96 percent used a card for paying in a hotel and 83 percent used their card in a restaurant, only 53 percent chose to use cards in a pub. In a crowded pub or bar, many guests are afraid that someone may see their PIN. In addition card transactions took longer than just placing cash on the counter. That was in 2013. Now, four years later, digital payments are up to 80 percent (Table 14.4). Customers are more comfortable with their digital payment systems, and businesses are getting better at taking digital payments. There are more terminals, most are mobile, and the transactions are performed faster than before.

Tap-to-pay systems based on cards were introduced in Norway in 2014. DNB has issued several million cards with tap-to-pay functionality, although there are still only a limited number of terminals that are capable of reading these cards. Of the customers who have a contactless card, 38 percent say that they use this function. The limit for avoiding the PIN is 200 kroner ($23), which is similar to the £20 limit that previously existed in the UK. However, it was found that the usability of tap-to-pay increased when the limit was extended to £30.

In fact, instead of having one limit for everyone, customers should perhaps have some say here. A cash user is free to decide how much to have in his or her wallet, so the same freedom should be allowed to customers who use digital systems. We shall return to this discussion in Chapter 15, showing that flexibility and security can go hand in hand.

Mobile Payments

	TOTAL	Age group				Sex	
		18-29	30-39	40-49	50 +	Male	Female
Yes	25 %	33 %	23 %	24 %	24 %	25 %	26 %
2015:	*22*	*17*	*24*	*24*	*23*	*25*	*19*
No	45 %	50 %	48 %	45 %	41 %	46 %	43 %
2015:	*43*	*56*	*43*	*41*	*38*	*40*	*47*
Maybe	12 %	13 %	14 %	14 %	11 %	14 %	11 %
2015:	*15*	*15*	*17*	*18*	*13*	*17*	*13*
Don't know	18 %	5 %	15 %	18 %	25 %	15 %	21 %
2015:	*20*	*12*	*15*	*17*	*26*	*17*	*22*
TOTAL	100 %	100 %	100 %	100 %	100 %	100 %	100 %

Table 14.5 Will mobile payments make it easier to pay in a pub or bar?

The surveys asked whether new solutions for paying with a smartphone will make it easier to pay at a pub or bar (Table 14.5). Only 25 percent believe this is the case, up from 22 percent in 2015. As many as 45 percent answered no, suggesting they are happy with their debit or credit cards, or—for a few—using cash.

	TOTAL	Age group				Sex	
		18-29	30-39	40-49	50 +	Male	Female
Yes	68 %	90 %	82 %	80 %	48 %	65 %	71 %
No	29 %	9 %	17 %	18 %	49 %	32 %	27 %
Maybe/don't know	3 %	2 %	2 %	3 %	4 %	3 %	3 %
TOTAL	100 %	100 %	100 %	100 %	100 %	100 %	100 %

Table 14.6 Have you used, or do you plan to use a smartphone for person-to-person payments?

When it comes to person-to-person payments, the responses were very different (Table 14.6). Sixty-eight percent of respondents currently use their smartphones for such payments today or plan to do so. This number rises to 90 percent in the youngest age group (18 to 29).

The Future

			Age group				Sex	
		TOTAL	18-29	30-39	40-49	50 +	Male	Female
Yes		38 %	37 %	30 %	33 %	44 %	43 %	33 %
	2015:	*34*	*33*	*29*	*30*	*38*	*34*	*35*
No		53 %	56 %	59 %	58 %	47 %	48 %	58 %
	2015:	*54*	*58*	*62*	*61*	*47*	*57*	*51*
Maybe		6 %	6 %	8 %	6 %	6 %	6 %	7 %
	2015:	*7*	*7*	*5*	*8*	*8*	*6*	*9*
Don't know		3 %	1 %	3 %	2 %	4 %	3 %	3 %
	2015:	*5*	*2*	*5*	*1*	*7*	*4*	*5*
TOTAL		100 %	100 %	100 %	100 %	100 %	100 %	100 %

Table 14.7 If you look ten years into the future, do you think
that we will still use cash?

Overall, 53 percent of respondents believe that cash will still be around in 2027 (Table 14.7). The numbers are consistent with the answers we received in 2015.

Summing up Norway

All new technologies have users who, for various reasons, stay behind in the old. While most consumers bought a color TV in the 1960s or 1970s, some kept their black-and-white sets until the compatible broadcasts were shut down. Even though mobile phones cover most of the population, there are still those that see the advantages of a fixed phone line. By offering courses on how to use Internet banking, a major Norwegian bank tried to transfer the small group of customers that use paper and telephone systems to the modern systems. As this did not work (customers did not use their new Internet accounts), they decided to maintain the opportunity to pay bills by post and to access accounts by the automatic phone systems.

Therefore, we should expect that cash and paper will be around for a long time in Norway. Still, if the trends that we have presented here continue, and there is absolutely no reason why they should not, cash will lose its importance in the payment system. Today, cash is artificially maintained by the fact that it is the default legal tender. While businesses are forced to take cash, many have challenged the regulations by discouraging the use of cash or by not accepting cash. As we have seen, Internet shopping, buying tickets, subscriptions, and so on will favor digital payments.

Customers who want to pay with cash will have difficulty doing so. As we have seen, banks are closing branch offices, removing cash handling from the remaining offices and reducing the number of ATMs. Many customers will see the futility of using a card to withdraw cash from an ATM and then using the cash at a store instead. Some believe that cash is more effective, at least for small payments, not understanding that cash is the most expensive option for the merchant. Over time, we should expect that most customers learn to use their digital payment systems directly at the merchants. Tap-to-pay solutions, either using a card or a smartphone, will be an important incentive, clearly showing the advantage of a digital payment for small amounts. Zero fees on digital payments, the norm in Norway today, and increasing fees for withdrawing cash will also be important elements in the transition to a cash-free society.

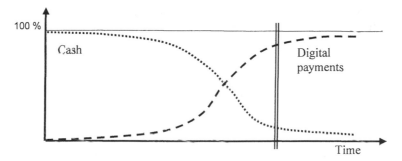

Figure 14.10. From cash to digital payments

We presented Figure 14.10 earlier in a general sense; here it is repeated showing the decline of cash and the advance of digital payments. There is a turning point where use of the new technology accelerates, with a similar strong decline in the previous technology. Norway is at this point with regard to cash, illustrated with the vertical line in the figure. As we have seen, the infrastructure for digital payments is in place. Digital payments are convenient, ubiquitous, and do not involve any fees for consumers. For many types of payments, including all online payments, a digital payment is the only option.

People who try to use cash will face many problems. As we have seen, it will be more difficult to get cash. In the future, there will probably be higher fees for using an ATM. More and more businesses will try to avoid handling cash as only very few customers choose this alternative. Even if the authorities want to enforce the "must accept cash" regulation, it will be increasingly impractical and expensive for businesses to receive change and return the cash to their bank accounts. Therefore, it may only be a matter of time before this regulation is removed.

Today, even Norwegian beggars accept digital payments. This is a necessity as most people no longer carry cash. No, they don't have a card reader. Instead mobile payments are used. A favored service is Vipps, a system in which the receiver is identified with his or her phone number (earlier presented in Figure 9.3). In my home town, the city of Molde, Norway, with 25,000 inhabitants, none of the banks offer cash services. They direct customers to the few ATMs that are left. The last cash-only business, a hairdresser, had also to accept digital payments by the end of 2017. With this, all businesses in the town accept digital payments.

A survey from 2018, performed by Finans Norway, shows a dramatic change in the last year, now eight out of ten pays digitally, independently of the amount.[10]

One would expect that authorities, such as the central bank, would be at the forefront of requiring digital payments and would try to get rid of cash that stimulates the underground economy, but they are not. Several institutions in Norway, from the union of finance employees to Økokrim (Norwegian National Authority for Investigation and Prosecution of Economic and Environmental Crime) have asked that the highest denomination bill of 1000 kroner (equivalent to approximately US$125) be removed, following the phase-out of the €500 note in the EU. In a letter to the Ministry of Finance, the central bank (Norges Bank) explained its position if the 1000 kroner bill is removed: "Alternatively, criminals could use the higher denominations from other countries, such as the Euro. Euro is readily available, and greater use of Euro for value retention or criminal activity at the expense of the Norwegian krone will represent a transfer of wealth from Norway to European countries. The cost may be substantial if the 1000 kroner note is removed."[11]

This is an interesting statement: the central bank seems to support criminal activity as long as it can make money out of it. Of course, the argument that if we don't do something others will is fairly common. If we don't sell guns to the rebels, others will; if we don't take the money that we found on the street, others will. At least some politicians have protested,[12] and one newspaper called it a "criminal defense for maintaining the 1000 kroner note."[13]

So why does the central bank go so far in upholding a criminal activity? The answer, as I have discussed previously, is seigniorage. This is a significant part of the income to Norges Bank and will disappear along with cash. As we have seen, cash has only a marginal place in the Norwegian economy. If the high-denomination banknotes (in Norway, that would be the 1000 kroner note and the 500 kroner note) are discontinued, the seigniorage for the rest of the notes and the coins would be marginal.

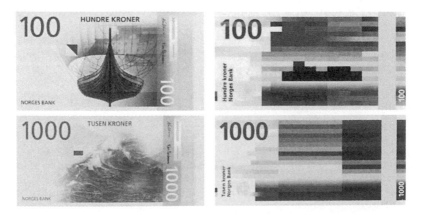

Figure 14.11. A new series of Norwegian bank notes (Norges Bank)

Norges Bank is introducing a new series of banknotes (Figure 14.11)[14] that retain the same denominations as previously: 50, 100, 200, 500 and 1000 kroner. It is a good idea to present these in the media, as most Norwegians may never own one of these bills, at least not the high denominations.

The focus that Norges Bank may have on cash may be dangerous in an economy that is going digital. As discussed earlier, a digital economy may be vulnerable for cyber attacks and for faults in power, mobile or data networks, and for errors in the payment systems. We currently experience all of these problems. After a heavy storm, there may be communities without power and where the mobile networks are not working. Norwegians regularly experience that payment systems or Internet banking have intermittent technical problems. Most of these problems can be avoided, but it will be expensive. Perhaps one cannot expect that banks, network providers, and the like will do this on their own initiative. That is, the country needs an authority that can define regulations for a digital economy; experience shows that one cannot rely on the market to provide secure systems.

As I have argued earlier, while we should avoid situations where authorities start fussing with the technical solutions, a better idea is to set levels of quality of service that should be required from systems, applying fines whenever these are broken. The participants—power distributors, mobile phone companies, broadband providers, banks, credit card companies, and so on—must then find the means to offer continuous service. In many cases this can be achieved by cooperating with other actors. Where cooperation is not an option it may be necessary to duplicate equipment.

In determining security, it is also important to take cybercrime into account. One way of doing this is to establish groups that look for weaknesses in existing banking and payment systems, so-called "white hat" hackers. Organizing this should also be an important task for the central bank. Security will be an even more important issue in the future. Even the innocent computers that are now becoming part of many products as part of the Internet of Things (IoT), in door locks, washing machines, temperature control, and so on can be taken over and used for computer attacks. One way to take down a website is to infect thousands, perhaps millions of computers, with malware and then use these to access a website. Such "denial-of-service" attacks can be used by competitors, political activists, or foreign governments. To avoid these attacks, there would have to be a wider agenda than just securing the systems that handle the digital economy.

Notes

[1] Kunderetta betalingsformidling 2016, Norges Bank Memo, no 2, 2017.
[2] http://www.frbsf.org/cash/publications/fed-notes/2016/november/state-of-cash-2015-diary-consumer-payment-choice
[3] The compound annual growth rate (CAGR) is a measure of growth of an investment.
[4] http://blog.euromonitor.com/2016/09/consumer-card-transactions-overtake-cash-payments-first-time-2016.html

[5]   MedieNorge, fakta om norske medier (in Norwegian)
      (http://www.medienorge.uib.no/statistikk/medium/ikt/379)
[6]   Kunderetta betalingsformidling 2015 , Norges Bank Memo, no 1, 2016.
[7]   Aftenposten, 3 February, 2016 (in Norwegian).
[8]   http://www.norstatgroup.com/
[9]   The following reports are in Norwegian:
      1. Kai A. Olsen and Kjetil Staalesen (2013) "Et Kontantfritt Reiseliv, Konsek-
         vensutredning," NHO Reiseliv.
      2. Kai A. Olsen (2015) "Et Kontantfritt Reiseliv—Norge Blir Kontantfritt"
         NHO Reiseliv.
      3. Kai A. Olsen (2017) "Norge Blir Kontantfritt," NHO and Molde University
         College.
[10] Finans Norge, Forbruker- og finanstrender 2018 (in Norwegian), 20. April,
      https://www.finansnorge.no/aktuelt/sporreundersokelser/forbruker-og-
      finanstrender/forbruker--og-finanstrender-2018/forbruker--og-finanstrender-2018/
[11] http://e24.no/makro-og-politikk/kriminelt-forsvar-for-1000-lappen/20022212 (my
      translation to English).
[12] http://e24.no/kommentarer/spaltister/tusenlappen-svart-men-viktig/20021813 (in
      Norwegian).
[13] http://e24.no/makro-og-politikk/kriminelt-forsvar-for-1000-lappen/20022212 (in
      Norwegian).
[14] http://www.norges-bank.no/sedler-og-mynter/ny-seddelserie/

Chapter 15
New Systems

In the first phase of implementing a computer system, the aims may be to improve usability and efficiency with regard to the previous system. This is also the case for digital payment systems. It is more convenient for customers to pay digitally; it simplifies accounting for the merchant, increases security, and is, overall, much more efficient than counting, storing, protecting, and transporting cash. The next phase will often be to try to improve the current systems, maybe with a focus on additional functionality, and perhaps also to offer completely new services.

We have also seen this development for mobile phones. The first phase offered a clone of the fixed-line telephone without the line. The functions were similar: calling and receiving. In the next phase, we got better user interfaces and functions such as texting. Then new technology, smaller and more powerful computers, made it possible to drop the initial idea of a phone. We then had a device that we could carry in our pocket with all the features we expect from a computer, with the phone functionality almost as an add-on. Today customers use their smartphones for surfing the Internet, checking email, going on Facebook, sending Snaps or Tweets, listening to music or reading books, paying for commuter tickets, and much more, such as making phone calls. The first phase of digital music was no big deal. A CD may have stored more data than a vinyl record or a cassette tape and may have been more robust, but the music was still on a physical media that had to be bought in a store or sent by mail. The revolution came with the second phase. With the Internet digital music could be stored on the net and streamed to customers.

Digital payment systems have completed the first phase. The customers and merchants have been offered basic functionality. Most importantly, in digital economies these services should be available everywhere. As we have seen, Norway is an example; there, 37 percent of people say that they never use cash. If we look at the latest data, eight out of ten always pays digitally. That is, one can operate without cash in this country. Norway and several other countries are now into the second phase—improving usability and efficiency. Tap-to-pay systems and the use of the smartphone for payments are good examples.

However, it should be possible to take several steps forward in this quest to find better and simpler user interfaces. I will discuss some options below. I shall also speculate on what the next phases can offer—that is, when we try to make use of all the data that is available in digital form. Here the focus may be on fulfilling user needs in a broader sense.

Better Overview

In the surveys I helped to conduct (see Chapter 14), we asked the people who used cash why they preferred to pay this way. Twenty-eight percent answered that it gave them better control. This is interesting; in most other cases users will say that the advantage of the digital system is that it offers a better overview and thus better control. However, the 28 percent have a good reason for giving this answer. With cash, a person can open his or her wallet to see what is left. The digital customer may have to access the bank account, and even then the balance may not be updated. There may be payments that are not yet registered. In the near future, we must expect that accounts will be immediately updated, but digital users will still have to log in to their account.

The solution is to integrate the bank account and the payments performed by using a debit or a credit card with a smartphone. Instead of discussing whether it is most convenient to pay with a card or the phone, the answer may be good integration. For example, with good network coverage the smartphone can be programmed to show the account balance and the last transactions at any time. Since this is a "view-only" mode, no login should be necessary except to unlock the phone. In fact, users should also be able to have this information on the home screen, making it available just by looking at the phone. When making a payment, by card or by phone, the transaction data could come up on this screen immediately, even if a third-party credit card is used. With the new EU regulations,[1] the customer may be able to let these data be available to others, in this case a smartphone app. There are already early implementations of such a system.

An additional advantage is that a system like this will improve security. Misuse of a card will most probably be detected immediately when the transaction comes up on the smartphone. That is, a system like this improves both usability and security.

Users who are on their way into town for a night out may have the ability to set limits. That is, instead of taking a given amount in cash to limit expenses, they may be able to set this limit with the phone app. The smartphone may then issue an alarm when the limit is reached, mimicking an empty wallet. In practice, this will never be a true limit on expenses, as the user will always be able to use more than the preset amount. However, even a user who takes a given amount in cash will most probably have a credit card in reserve.

Simpler Payments

Tap-to-pay solutions have been a success. Paying for a cup of coffee or a beer can now be done just by tapping the terminal with the card or the smartphone, avoiding a PIN. While the limit, often around $25 to $60, may be a nuisance, taking the PIN away from most transactions simplifies the payment and increases security. In practice, PINs should only be used for the higher amounts. While someone could steal your card and use tap-to-pay functionality to buy beer in a pub, misuse of the card will be limited without the PIN.

Ideally, users should be able to customize the payment system. For example, many of my own digital payments are made at the grocer, fishmonger, gas station, and the cafeteria at work. What I would like to do is to tell the credit card companies that payments to these merchants should not require a PIN. I would then be required to set a maximum transaction amount and perhaps a per-day and per-week maximum. Clearly this will simplify payments. One can now tap-to-pay (nearly) independently of the amount in all the stores that we visit frequently.

Improved Security

Implementing the PIN-free transactions described above will also improve security. Some criminals will stand in line behind a customer and note their PIN, then steal the card and withdraw the maximum possible amount from an ATM. With PIN-free transactions at selected merchants, there will be no PIN to register. Of course, a thief who steals a card may go back into the fishmonger and try to buy a kilo of cod. This will not be easy because the thief will not know the limit and the customer may soon be back into the store to look for his card. In any case, the priority of a thief is probably not to get fish. At a grocery store, the thief may be able to get products, such as alcohol and cigarettes, that may be possible to sell, but the risks will be very high, especially compared to the possible gains.

If this system is combined by offering all transactions on the smartphone, as we discussed above, the customer will be warned immediately of any transaction, making it even more risky for the thief.

Many customers do not use ATMs, but it isn't possible to block a card for such transactions. One must take the unnecessary risk that someone will steal their card and withdraw cash. The solution is to customize safety measures. For example, people who never use an ATM should be able to block the card for these transactions, or, if they want to retain the possibility of withdrawing cash, add additional security measures. One option is to have an additional PIN code or perhaps use a smartphone to remove the block.

Automatic Data Retrieval for the Authorities

While taxpayers in the United States often have to hire specialists to complete the forms that the IRS (Internal Revenue Service) demands each year, in Norway the tax authorities will complete the forms automatically for most taxpayers. While it is recommended that you check the form, ordinary taxpayers will normally accept the form as it is. There is not even a requirement to sign and submit the form. In 2017 the tax office marked the change by replacing the name of the form, from tax declaration ("selvangivelse") to tax message ("skattemelding"). The automation of this task is made possible by the fact that all data on individuals—for income, bank accounts, stocks, real estate, and so on—are available to the authorities online. This means there is no need for ordinary taxpayers to spend any time on this. A complicated and tedious task has been automated. At the same time, extensive controls of the ordinary taxpayer are no longer necessary and the tax authorities are quite confident that they have included everything.

Most of the data for small business owners is also available in the online form. The owner's job is to insert the few data items that are not available for the tax administration. Every year, more and more of the necessary data will be made available to the tax office in digital form, making this task simpler and simpler.

The advantage for the tax office is that much of its work, such as computing the taxes that everybody has to pay, can now be performed by the computer. The advantage for the taxpayers is that the whole process can be performed in much less time than before. Tax authorities can then use their time on the more complex cases.

This idea of retrieving data that are available online can be extended to many other areas. While businesses complain that they have to cope with an unlimited demand for data, such as statistics, these data could also be retrieved automatically. The business would need a set of web services[2] that the authorities could access in order to retrieve the data from the computer system of each organization. This is not difficult to implement. Not only would this remove the hassle of collecting the data, but authorities would also have the ability to present updated data. For example, if the number of employees was available digitally, this number could be updated every night for the whole country. Statistics could then be offered on a daily basis.

In this book I have concentrated on economic data, but such a system could be extended to any type of data. For example, when medical records are digitalized, health authorities could produce updated statistics, making it easier to detect epidemics early. This can be done with aggregated data, so there will be no need to have any personal identifiers.

Future Banking

Some Internet banks have tried to create additional profits by selling other services, ranging from investment offers to travel packages, on their websites. This is viewed as an important business model for retail banks, as the possibilities of generating revenue on the traditional banking services decrease with increased automation. However, the model may not be valid because Internet banks are used in a different manner than traditional banks. In an all-electronic world, it is so much simpler to set up an automatic bill payment system. For example, instead of having to pay a bill on the due date, the Internet bank can be told to perform this transaction automatically if the customer does not interfere. Customers can still check their bills by going through the data provided in an email, but in the normal situation, the customer will have no need to access the banking portal. Already, with the services that an Internet bank offers for automatic payments, most personal transactions can be handled automatically. The day of the complete automatic banking system may not be very far off, even if the odd bill has to be entered manually.

Banks are conservative institutions. Even Internet banks that rely on new technology for their services have not implemented many of the new functions that online information makes possible—that is, to work as "information banks," providing new information-based services for their customers as a replacement for the old cash-based services. These could be as simple as organizing and printing existing information or, more dramatically, using this information to act on the customer's behalf.

For example, while businesses are required to have full accounting, few consumers are willing to do the necessary work, even if most of us would be interested to see an account of last year's income and expenditures. If we agree to use the bank account and accompanying credit or debit cards for a substantial part of our expenses, or allow the bank to collect these data from other parties, and accepting that the bank retrieve information on what we bought from point-of-sale terminals and bills, such a full account could be made automatically. The same data also could be used as a basis for the next year's budget, perhaps provided by the bank in the form of a spreadsheet so that we could adjust the numbers ourselves.

With additional data, a good banking service—or perhaps we should call it the automatic financial advisor—should be able to offer recommendations how we could save money next year. The ingredients for such a service will be data from all transactions that the customer has performed, data from other customers, and from the outside environment. This can be used as input to smart algorithms that use artificial intelligence techniques to withdraw information. Such a system could offer advice to customers, such as:

- Save $234 per year by changing utility companies.
- Change to a different account type and receive higher interest rates.
- A customer card would receive a $56 rebate on five overnight stays at Hilton hotels last month.
- Credit card interest amounted to $1260, with an average rate of 17 percent. Consider a second mortgage loan instead, with savings of up to $800.
- You had three visits to the Science Center, with a total of $190 in entrance fees. Consider a family membership that gives unlimited access for $110, of which $98 is tax-deductible.

The bank could take an even more active role. For example, we could instruct our bank to find the best deal for utilities, let the bank automatically order preferred customer's rebates on travel, and bargain on our behalf for goods that we need. The banks will be in a good position to offer such services since they have the necessary data and the contact points both with customers and companies. But it is important that they are trusted.

To offer such services the bank needs information that will raise privacy issues. However, where we have earlier trusted banks with our money, we must now trust that the bank will guard this information and not let it be used to our disadvantage. Today, we expect that the bank will compute interest rates correctly. Few people take the time to check these calculations. In the same way, we may trust the bank to provide the services mentioned above in the future—for example, expecting that the bank will negotiate the best utility contract.

As the standard retail banking functions are replaced by a computer, making it more difficult to get high revenues on these functions, banks will need these additional services to survive in the retail market. The Internet-based banks of the future will still be in the trust business, but only parts of the data on their disks may represent money.

There are already companies that offer software-assisted wealth management for ordinary citizens. With smarter software and access to all relevant data, it should be possible to implement the automated personal assistant described above, a "robot" that could analyze our economy, using data mining methods, and then offer advice based on this. With the new EU regulations these institutions will be able to gain access to data from our bank and credit card accounts (with our permission).[3] Thus, if the banks do not provide these new services, others will.

Conclusion

In this chapter we have taken a peek into the future and presented some of the services that are possible. There may be many more, but central to most will be that they are able to use the data that are generated from all the underlying systems in novel ways. The advantage is that financial data are formalized and can therefore be analyzed using novel techniques. Machine learning and other AI methods will offer the best results when the inputs are formalized.

Notes

[1] The Payments Services Directive (PSD2).

[2] A web service is an application that can be invoked over the net. Each company can then offer access to such a service from government offices. The web service will be connected to the back-office systems. When the service is invoked it will retrieve the necessary data from the back-office systems and return these to the calling agency.

[3] The Payments Services Directive (PSD2).

Chapter 16
The Cash-Free society

As far back as the 1970s, there were prophesies about a paperless society. Even today, nearly fifty years later, I see paper when I look around my office. There is one very good reason: it is easier to read a document on paper than on any other media. When it is necessary to read long reports or scientific articles, I prefer to have these on paper. Paper not only has higher print quality, but it is also more convenient for scanning and navigating a document. For example, paper is not limited to screen size; one may look at many pages at the same time. This is also the reason why I still prefer to read newspapers in the traditional paper form, using the digital versions mostly for getting the latest news.

However, the use of paper has been significantly reduced, both in my office and in most others. All my teaching material is online. Email has replaced letters. I no longer check the (physical) mailbox every day. As a colleague told me, if you send a letter, remember to give a notification by email so that the recipient will check his or her mailbox. Programming—a task that previously produced a lot of paper—is now being performed online. Online tools make the process significantly simpler. Writing falls into the same category.

I prefer reading books on paper, but my Kindle makes it possible to buy a book in just a few button clicks. The electronic ink enables convenient reading since it uses the light in the room, similar to paper. As long as the book is mainly text, the resolution, format, and black-and-white presentation are acceptable. It is not suitable for complex tables, figures, and photos. Kindle's electronic ink has been around for many years. While we might expect to now be offered high resolution, full color, and letter format readers, evolution does not always go as fast as we expect. All in all, we should expect to see a decline in the use of paper, especially if we get a good document viewer. However, the transition is quite slow and we may have paper for books, magazines, and reports far into the future.

So, we should be careful about prophesies of a cash-free society. As with paper, cash may be here for a long time. However, there are some differences. While we still do not have a great digital viewer, we have all the technology that we need to pay digitally. Digital payments have many advantages compared to cash and, with the right infrastructure in place, will be performed more efficiently and more conveniently than using cash. Unlike cash, digital payments can be applied in all situations, in stores, as well as online.

All operations are formalized and data requirements are modest, with no reduction in quality. In fact, the opposite applies: digital payments can be performed faster, especially if one can avoid using a PIN. Digital payments are also more secure and clearly more efficient and environmentally friendly since people can avoid transporting the cash. In addition, the advantage of getting all the transaction data in digital form is quite high as the data can be used for other purposes as well.

Earlier, I presented Norway as an example of a country that will, in practice, be cash-free within a few years. However, I could have presented several modern countries where cash is still king. As we have seen, around 80 percent of all transactions are performed in cash in Germany, and in Belgium more than 60 percent are in cash.

This shows that things take time. In order to go cash-free it is necessary to have an excellent technological infrastructure with full-coverage broadband networks. Banks and other companies need to provide good digital services, preferably without fees. Trust in the new systems is also important. Government guarantees of savings are then important. Finally, one must overcome tradition. Many people feel comfortable with cash, and many see the advantages of being able to generate income in a form that the tax collector cannot see.

In many situations, customers have the option to pay digitally or with cash. As the technology advances there will be many situations where digital payments are the only practical option. We have seen how the payment task is integrated in the overall process of buying goods, subscriptions, tickets, and many other processes online. For net commerce, cash is not an option; in fact, this will be the case for all computer-based transactions where there is no operator available to take cash. The move from traditional shopping to online shopping will also make cash less useful. As more and more payment functions are automated, the option to use cash may disappear. In addition, it may not be easy to get cash when banks close down cash handling and reduce the number of ATMs.

Therefore, we should expect that all countries will move toward digital payments, but with a transition speed that varies from country to country. International pressure, for example from large international companies that can offer excellent digital services, will act as a pressure for local banks; if they don't offer the same services, they may lose customers. While they may retain the older generation that use cash as customers, they may lose the younger customers who are comfortable with using the Internet and expect to receive most services for free.

Norway is an example of a country where the transition to a digital economy has made great strides. Nearly every merchant in Norway will accept digital payment, many will try to avoid cash payments, and a few will even decline to take cash. The transition to digital payments is supported by a no-charge policy for customers. Credit cards are free; there may be a yearly fee for a debit card but there are no transaction fees. As we have seen, there is no lower limit for digital payments, which enables a large part of the population to use digital payments only. Combined with tap-to-pay solutions digital payments mean there is no need to retain cash.

Fees for merchants may range from 1.5–3 percent for credit cards, while some cards may even require as much as 5 percent. The fee for debit cards is often very low, perhaps only a few cents. Until now most merchants have accepted these fees, but in some countries the fee will be added to the bill. Ideally, to get the most efficient economy, the customer should cover the expenses. For a $100 amount, a customer who uses a debit card should pay $100, $103 with a credit card, and probably $105 for cash. To avoid excessive charges, it is important that there is one low-cost alternative; today this is cash but it should be a debit card.

While merchants still have to handle cash, the amounts are very low. As the owner of my local grocery store told me, his cash share is now at 5 percent. This amount drops by one percentage point every year, making him practically cash-free in a few years. The cash percentage was so low that he skipped plans to install a secure cash register—that is, an "inverse ATM" where bills and coins are offered to the machine that will give change automatically. Still, there will probably always be customers who want to pay with cash, but some day the store may say no—that is, when the cost of handling cash is larger than the profits on these customers.

Norway, like the other Scandinavian countries, has a high tax rate. For personal income it may be as high as 50 percent. There is also a value-added tax that ranges from 10 to 25 percent. This offers an incentive for avoiding tax. With income in cash there are several possibilities both to hide income and to pay employees off the record. This is much more difficult when customers pay digitally, since digital transactions can be traced. While there are a few small businesses that only take cash, these are now having problems because many of their customers do not have cash and the nearest ATM may be far away. It is also questionable as to how long they may avoid taking digital payments. For example, all London black cabs are now required to accept card payments.[1]

While digital payments make it difficult for national businesses to avoid paying their taxes, they cannot help with the tax evasion procedures that many large international companies use. Such companies typically move money to tax havens before taxation, which gives them an unfair advantage over the locals. For example, a small local coffee shop must pay all its taxes, especially when customers pay digitally, while a large international competitor can avoid paying taxes. Politicians in many countries are now trying to establish regulations that require companies to pay the tax in the country where the income originates.[2]

Cash is anonymous, as are some forms of digital payments. However, we cannot offer full anonymity if we want to reap the advantages of a digital economy. As we have seen, customers do not seem to be very concerned. Only a tiny percentage of people offer anonymity as the reason why they use cash. In many countries, most customers have a smartphone; they may be active on social networks, use email and many other systems that do not offer full privacy. Most users do not seem to be very concerned about this situation. Privacy is an interesting concept that is very much in the media focus today. However, if we go back a hundred years or more, when people lived in small cities and communities, there was not much privacy. This had advantages and disadvantages. One could receive help from friends and family if there was a problem, but informal surveillance could be extensive. There was little room for anonymity, although larger cities offered the possibility of more privacy. We then came to expect that this was the norm.

Today, with many forms of digitalization, from social media to smartphones and digital payments, there is a threat to privacy, although there are laws to prevent the data from being misused. These laws are designed to avoid situations such as an insurance company monitoring payments and perhaps refusing to sell someone life insurance because he or she has been buying too many cigarettes. However, we may accept that they offer a discount if we agree to install a monitor in the car that supervises driving—for example, by noticing if we respect speed limits. This is now an option and in the future it may be a requirement. Perhaps this is a breach of privacy, but if it results in fewer accidents it may be an advantage for everyone.

Bitcoin and other cryptocurrencies may offer anonymity, at least to some extent. This anonymity is important for criminals. A typical scheme is to encourage users to open malicious attachments to emails, or find weaknesses in a computer's firewall, and then cryptograph all data so that it becomes inaccessible. The criminals will then ask for a ransom—in bitcoin. The WannaCry ransomware attack in May 2017 showed how vulnerable many systems are. While an updated Windows operating system was able to stop the attack, many institutions were caught with their pants down. FedEx, Renault, Telefonica in Spain, universities in China, German railroads, and Britain's National Health Service (NHS) all encountered major problems. None of these had updated systems, and the NHS even used Windows XP, an old version of Windows that Microsoft no longer supports. All in all, WannaCry is estimated to have taken down 200,000 computers around the world. The possibility of anonymous payments is an incentive for schemes such as this. As we have seen, there is a price for anonymity. To fight crime it seems reasonable to start by taking away one of the most important tools that criminals have—the anonymous payment. While this may stop most of these attacks, there may be other incentives to inflict harm on computer systems. Thus, updating software and having good security routines in place is always essential.

Many countries are on a path to a cash-free society. This has been a market-driven process. Customers and merchants have seen the advantages of digital payments, and technological companies, banks, credit card companies, and other financial institutions have provided the technology and the services. It seems that countries will be cash-free whether they want to or not. The disadvantage of a market-driven approach is that there may be issues that the market is not willing to solve. The two major issues are:

1. Offering digital payment opportunities to everyone, including customers with payment problems.

2. Requiring that we have secure systems that have 100 percent availability.

In some countries, such as Norway, it seems that the market may offer accounts and debit cards to everyone, but a requirement for banks to offer such a service will guarantee that everyone is included. As an alternative, the central bank can issue these debit cards. Thus, point 1 above is easy to solve.

Point 2 is more difficult. Experience shows that the systems that we have today are not secure enough. Customers experience blackouts, downtime for mobile networks, and problems with payment systems. Establishing secure systems that are available all the time and are protected for everything from hurricanes to crypto-criminals, are paramount. It is important to do this now, since many countries already have a digital and almost cash-free economy in practice. This is not something that can be left until the day when the last bill is revoked.

Thus, to get a smooth and secure transition from cash to a digital system, the authorities must take action. I suggest the following plan:

1. Penalize providers for any downtime in power networks, data networks, mobile phone networks, and payment systems. Such penalties need to be so high that companies have the incentives necessary to install backup solutions.
2. Offer regulations that allow technical cooperation between competitors in a crisis.
3. Require all merchants to accept digital payments.
4. Let merchants choose if they want to also accept cash.

5. Revoke the highest denomination bills, starting at the top and moving downward. This process can continue until the last coin is revoked—that is, when the country is cash-free or until one sees that the advantages of going further are low.

There may be arguments for discontinuing the process of revoking cash when only small denomination notes are left. However, there are clear advantages in moving to a fully digital economy. There will then be no need to retain systems for offering cash, such as ATMs, or to convert cash to digital. There will be no need to print notes, mint coins, or to transport and store cash. When a digital system is in place it can also handle small transactions efficiently.

As we have seen, the problem is getting the central banks to play an active part. Until now they have chosen to protect their own interests—that is, to protect the income they have from seigniorage. Clearly, the process toward a cash-free society will continue as a market-driven effort. But this may lead to a situation where a severe disruption of digital services is a possibility. This is already a risk in some countries, such as in Scandinavia. Cash will not be a practical backup in such a situation. In the worst case, there may be an absence of effective payment systems for a prolonged time. However, with secure systems in place, cash-free can be an aid to solving many of the problems that we face today.

Notes

[1] https://www.theguardian.com/uk-news/2016/feb/03/all-london-black-cabs-to-take-card-payments-from-october

[2] https://www.theguardian.com/business/2017/sep/21/tech-firms-tax-eu-turnover-google-amazon-apple

Index

account number, 73

advantages, 101

airport express train, 83

Alexander the Great, 45

Amazon, 22

American Airlines, 11

AML, 96

analog, 57

anonymity, 75, 96, 107, 111, 116

Ant Financial, 98

app, 21, 72, 83, 105

app for plumbers, 20

Apple Pay, 98

Asian financial crisis, 66

ATM, 49, 50, 78, 129, 136, 137, 143

auto manufacturer, 20

automated payment, 127

automation, 9, 10, 19, 35, 36

autonomous buses, 31

autonomous car, 5, 28

babysitter, 115

backup, 118

baht, 66

bank, 9, 50, 67, 87

bank account, 112

bank guarantee, 87

bank infrastructure, 90

Bank of England, 57, 66

bank office, 129

banknote, 46

bar code, 16, 18, 36, 60, 62

better overview, 141

big data, 26, 31

biometric, 15

bitcoin, 76, 95, 149

Bjerke, Rune, 128

black-market, 103

blockchain, 94, 95

booking, 2, 19

bot, 26

branch office, 128

Brin, Sergey, 28

CAGR, 125

card reader, 114

Carr, Nicolas, 13

cash, 48

cash as backup, 118

cash card, 69, 74

cash register, 50, 62

cash-free society, 137, 147

central bank, 65

checks, 4, 57

China, 46, 47, 98, 99

Christian, Brian, 26

clearing system, 73

coin, 45

commodity, 43

commuter ticket, 82

compact disk, 1

complex computer application, 25

compound annual growth rat, 125

computer network, 73

continuous service, 77

Corsera, 23

counterfeit, 47

creative, 13

credit card, 58, 68, 73, 114, 115, 127

crypto-criminal, 150

cryptocurrency, 95

currency, 53, 65

Curse of Cash, 94, 107

cybercrime, 139

data retrieval, 143

debit card, 68, 74, 127

digital currency, 93, 149

digital economy, 65

digital infrastructure, 71

digital payment, 4, 5, 81, 101, 111, 127, 149

digital security, 118

digitalization, 20

disadvantages, 111

discount structure, 2

disruptive, 3

DNB, 128

driverless car, 28

drone, 37

EAN, 15, 60

electronic invoice, 127

embedded chip, 81

environmental issues, 106

equipment, 62

exchange rate, 55

Facebook Credit, 98

fax machine, 78

Feige, Edgar L., 117

fighting crime, 101

financial crisis, 107

fingerprint, 15

fingerprint reader, 81

formalization, 4, 9, 10, 11, 18, 29, 88, 89

foundry, 12

fraud, 102

future banking, 144

GDPR, 82

General Data Protection Regulation, 82

Germany, 104, 124

gold, 65

gold guinea, 45

gold standard, 47

Google's translator, 27

Google+, 23

grammar checker, 26

grey zone, 115

guarantee, 67

hacker, 76

handling cash, 105

hiding funds, 104

high-denomination bill, 120

hikers, 32

home freezer, 18

HTML, 10, 19

HTTP, 10, 19

hype, 36

Iceland, 124

ICOs, 93

identification, 14

Industrial Revolution, 105

inflation, 47, 56

infrastructure, 71

initial coin offering, 93

integration, 15

Intel, 62

interface, 22

intermediate system, 78

internet, 89

internet bank, 87, 88, 127, 144

Internet of Things, 139

inventory control, 18

invoice, 127

ISBN, 60

Island dollar, 53

job market, 34

Kindle, 17, 147

Knowledge Navigator, 23

Kodak, 2, 16

KYC, 96

library, 17

login, 87

Loon, 37

loyalty cards, 31

Lydia, 45

M1, 124, 131

machine learning, 33

magnetic strip, 81

malware, 88

maps, 30

Microsoft Access, 13

minimum amount, 68

minimum fee, 68

mining, 95

Ministry of Finance, 138

mobile payment, 114, 135

money, 43, 46, 76, 108

M-Pesa, 79

music industry, 17

Nakamoto, Satoshi, 95

National Security Agency, 104

natural language, 25

net shopping, 84

network effect, 4, 15

new systems, 141

NFC, 72, 85

Norges Bank, 123, 138

Norway, 123

Norwegian, 2

Norwegian restaurant employees, 105

NSA, 104

Office Assistant, 23

off-the-shelf, 14

one-click, 22

Panama papers, 104

payment, 44, 81

personal assistant, 22

personal freedom, 108

personal service, 90

person-to-person, 85, 96, 114, 126

PIN, 15, 50, 74, 75, 81, 84, 87, 113, 142, 143, 147

pirates, 1

point of sale, 71

Polo, Marco, 46

power grid, 118

privacy, 5, 101, 108, 111, 145, 149

proofreading, 26, 27

pub or bar, 134

QR, 16, 61, 72, 84

real-time clearing, 85

record industry, 1

reduced cost, 104

redundancy, 119

representation, 16

reward point, 97

RFID, 61, 71, 81

Riksbank, 93

robot, 10, 34

Rogoff, Kenneth, 94, 107

RSCoin, 94

Ruter, 83

Ryanair, 2

SABRE, 11

sale of music, 1

Scandinavia, 149

security, 75, 87, 119, 139, 143

seigniorage, 54, 116

senior citizen, 113

Simon, Herbert, 25

smartphone, 1, 22, 71, 81, 84, 142

social security, 101

Soros, George, 66

standard, 90

Stockholms Banco, 46

survey, 131

Swiss franc, 66

Swiss National Bank, 66

tax avoidance, 103

tax evation, 107

Tay, 27

terminal, 71, 126

Thailand, 66

ticket app, 72

time of arrival, 33

tourist, 69, 114

transaction fee, 67

translation, 25

translation, 26

travel regulation, 20

trust, 65, 67, 87

two-factor authentication, 81

U.S. Treasury, 117

uncle Joe, 53

United States, 4, 124

UPC, 60

usability, 113

VAT, 108

virtual national currency, 93

WannaCry, 150

wealth, 43, 44

wealth tax, 107

WeChat, 98, 99

wristwatch, 22

zero interest bound, 106

About the Author

Kai A. Olsen is a professor of informatics (computing science) at Molde University College, University of Bergen and at Oslo Metropolitan University, Norway. He is an adjunct professor at the School of Computing and Information, University of Pittsburgh. Olsen's main research interests are IT strategy and human-computer interaction (HCI). He has been a pioneer in developing software systems for PCs, information systems for primary health care, and systems for visualization. He has written several books and has published more than a hundred scientific papers, in addition to numerous articles in Norwegian newspapers. He acts as a consultant for Norwegian and US organizations.